Photographic Atlas of Entomology
and
Guide To Insect Identification

By
James L. Castner

Department of Biology
Pittsburg State University
Pittsburg, Kansas   66762

Published by

Feline Press
P.O. Box 357219
Gainesville, FL 32635   USA
e-mail: jlcastner@biologicalphotography.com

Photographic Atlas of Entomology
and
Guide To Insect Identification

By
James L. Castner

Published by:
Feline Press
P.O. Box 357219
Gainesville, FL
32635   USA

© 2000 by Feline Press, Inc.
First Printing 2000
Printed in China
ISBN 0-9625150-4-3

Library of Congress
Catalog Card Number:
00-134617

Preface

I created this photo atlas and identification guide in order to make it easier to recognize and learn about insects. It was designed for use by undergraduate and graduate level students who may be taking their first college courses in entomology. A student's initial exposure to entomology often involves an intimidating barrage of Latin nomenclature coupled with the seemingly insurmountable task of distinguishing between hundreds of insect specimens which all look alike. It is my sincere hope that the information presented in this book shall make the student's life somewhat easier. I have used my expertise as an entomologist and a photographer, combined with my experience in college teaching to produce a work that is easy to use and covers the basics. While the coverage in a given course may vary from professor to professor, and with geographical location, I have tried to provide a foundation for any general entomology or insect taxonomy courses given throughout the country.

The use of this guide provides the potential for a greater degree of consistency in how an introductory entomology laboratory or insect taxonomy course is taught. It is common practice in many universities for the laboratory sections of courses to be taught by graduate students or teaching assistants. One lab section of the same course may therefore vary greatly from the next, resulting in a difference in the quality and exposure students receive to the subject matter. This guide will assure that certain information will be made available to each student.

The images featured in this guide are all from actual photographs and/or digital scans. This should provide the student with an illustration that is much more likely to be similar in appearance to the actual specimen under the scope than would line drawings, as are so often used in similar manuals. I tried to use photographs of live insects whenever possible, but there are also many shots of pinned insects as well. This is not necessarily a drawback however, as students will most likely be looking at preserved specimens in the lab and collecting live specimens to make their own collection.

I have eliminated the original background in almost all illustrations, replacing it with a muted neutral gray so that the subject stands out more clearly. In doing so, however, I sacrificed some of the anatomical detail in some cases, while creating some unusual artifacts in others. For example, the wings of the papaya fruit fly pictured in Figure 520 appear green. This is actually the original background which was a green leaf. Sometimes pinned insects were originally photographed on a blue background. This resulted in blue-tinted wings in some cases. Extremely spiny, hairy, or fuzzy insects suffered the most from my handiwork. Unless such structures were important recognition features, I often eliminated them while tracing the outline of the insect. A good example would be the numerous spines found on the legs of most roaches. This was done to save time and make this project feasible. As it was, this guide took years to complete and features over 25 years worth of photographs in the 670 images.

I have tried to make the student's life easier with this photo atlas. I am open to suggestions on how to improve it, and would like to be informed of any errors. Students and teachers may e-mail me at their convenience.

Sincerely,

James L. Castner
jlcastner@biologicalphotography.com

Acknowledgments

Many people contributed significantly to this work by generously providing information, specimens for photography, or reviewing sections of the text. The curators, staff, and research associates of the Florida State Collection of Arthropods (Gainesville, FL) were particularly gracious and helpful. From this group I would especially like to thank Julietta Brambila, G.B. Edwards, Gay Fortier, Avas Hamon, Bill Mauffray, Frank Mead, Paul Skelley, Lionel Stange, Gary Steck, Mike Thomas, and Jim Wiley. The personnel of the USDA/ARS Center for Medical, Agricultural and Veterinary Entomology Laboratory (Gainesville, FL) were equally helpful and I would like to thank in particular Lloyd Davis, Darryl Hall, Jerry Hogsette, Johnny Jackson, and Nancy Lowman. Other colleagues and friends from institutions across the country who also helped in a variety of ways are Jerry Bowman, Jason Byrd, Sid Dunkle, Howard Frank, Elmer Gray, Dave Nickle, and Rudi Scheffrahn.

I would especially like to thank Don Hall and Frank Slansky for taking the time to review this manuscript and for making many helpful suggestions which improved its clarity and usefulness.

I pestered and annoyed many people at all hours with questions pertaining to computers, graphics, digitizing, printing, and other aspects of the publication process. For sharing their expertise and remaining patient I would like to thank Jason Byrd, Jane Medley, Nancy Heinonen, Shari Hennes, Mark and Maria Minno, Steve Orf, Pat Payne, Milt Putnam, Joe Rackowski, George Venable, and Laurie Walz (the Digital Diva). As always, the efficient and friendly staff at the University of Florida's Office of Instructional Resources provided excellent service. From UF/OIR I would like to thank J.R. Hermsdorfer, John Knaub, Pat Payne, Kimberly Sanford, Rachel Stevens, and Claire Tingling. Its always a pleasure dealing with you.

Special thanks to Dan and Zane Greathouse and the helpful staff at the Greathouse Butterfly Farm in Earleton, FL. I thoroughly enjoyed the many trips to your educational facility where I've spent many a happy hour photographing butterflies and other insects. Your cooperation is really appreciated. I'd also like to thank Synthia Penn of Beneficial Insectary, Mike Cherim of The Green Spot, Marta Martinez Wells, and L. Tedders for helping me to procure live beneficials.

Teachers seldom get the credit that they deserve. Over the years many individuals have contributed to my development as an entomologist, biologist, and educator. First and foremost are my family members. My parents, my Uncle Pit and my Uncle Bob provided the best atmosphere and circumstances any child could ask for to learn about nature. As an undergraduate student at Rutgers University, teachers such as Wayne Crans, Bob Denno, and Elton Hansens made lasting impressions. During my graduate years at the University of Florida, I was fortunate to be exposed to the teaching techniques and expertise of Don Hall, Jack Jenkins, Jim Lloyd, Frank Slansky, and John Strayer. For learning about life, nothing can surpass the teachings of Freddie A. Johnson, where wisdom and humor go hand in hand. Thank you to everyone.

Photo Credits

I have endeavored to take my own photographs wherever possible in this work. However, in some instances, I have used with permission, images from other photographers, scientists, and institutions when they were available. These images can be identified by a credit line that is set in italics.

Certain colleagues have had a lifetime passion for certain insect taxa and this is reflected in the high quality images which they generously supplied. I would like to thank Sid Dunkle for his dragonfly photos. Its always nice to get material from a world expert. Bill Stark kindly supplied excellent stonefly and caddisfly photos on very short notice. Illustrations for nearly the entire section on the Odonata were provided by Forrest Mitchell and James Lasswell and their colleagues at the Texas Agricultural Experiment Station (TAES). They generously provided scanned images promptly, and with more cooperation than I could have hoped for. Many of these shots can be viewed on their wonderful 'digital dragonflies' website (www.dragonflies.org). I also appreciate the work of Charlie Warwick and gratefully acknowledge the talented staff of scientists at the Illinois Natural History Survey (INHS) for making photos available to me. Thanks for your help.

Closer to home, I would like to acknowledge those people at the Division of Plant Industry of the Florida Department of Agriculture and Consumer Services (FDACS/DPI). In particular, I thank Director Richard Gaskalla, Maeve McConnell, and photographer Jeff Lotz whose wonderful images have graced DPI publications for years. I would also like to thank the Institute of Food and Agricultural Sciences of the University of Florida (IFAS/UF) for permitting me to use photos that I myself took while employed by the Department of Entomology and Nematology. Used with their permission were the photos in the following figures: 47, 261, 270, 271, 336, 362, 382, 521, 541, 597. Many of the photos used to illustrate the section on termites were generously provided by termite expert Rudi Scheffrahn of the IFAS Ft. Lauderdale Research and Education Center.

The Agricultural Research Service of the United States Department of Agriculture (ARS/USDA) employs some extremely talented photographers and researchers. Their work is often used in government publications, contributing significantly to the understanding of the topic being illustrated and discussed. In particular, I would like to thank Scott Bauer, Jack Dykinga, and John Kucharski.

Dedication

It has been nearly 27 years since a kindly professor of medical entomology at Rutgers University recognized that I had an interest in learning about insect photography. Fascinated by the close-up shots of bloodsucking insects that adorned his office, I would often stop on my way to class to look at the human blood swirling into the gut of a body louse, visible through its nearly transparent skin. Equally of interest was a shot of a conenose bug, swollen like a little balloon as it engorged off its host. After several conversations in the hallway, I was invited to spend a full evening at this professor's home, an outing that truly changed my life. I was tutored for hours in the use of a bellows, in macro lenses, and in all the tricks of dealing with live insects as subjects. That evening was to have long-reaching effects in my decision to pursue graduate training in entomology, and in fueling my desire to become an accomplished insect photographer.

I would like to sincerely thank Dr. Elton J. Hansens, retired Professor of Medical Entomology, my advisor while attending Rutgers University. Your time and efforts on my behalf, and your many other students, are greatly appreciated.

James L. Castner

Table of Contents

Table of Contents

Table of Contents

Table of Contents

Table of Contents

Table of Contents

Table of Contents

Table of Contents

Table of Contents

Table of Contents

Table of Contents

Table of Contents

Introduction

There are an estimated 30 million species of insects in the world! Fortunately, only about one million of them have been named and you as an entomology student will not be asked to learn to identify every one of them. However, distinguishing between and recognizing the taxa of arthropods and insects included in your course may seem like an equally impossible task. The purpose of this guide is to enable you to learn the basics of insect identification as easily (and painlessly) as possible. Before each ID quiz I gave, the hallway outside the lab was lined with students frenetically cramming and reviewing their notes. It is my sincere hope that the highly organized and illustrated nature of this guide and photo atlas will take some of the anxiety away from students and allow them to enjoy the experience a little bit more.

An indispensable tool in learning to separate insects to their various taxonomic groups is the dichotomous key. The use of such keys will be covered in the next section. Your instructor may provide you with customized keys that deal with the specific orders and families that you will discuss in your course, and that you are most likely to find in your particular geographical region. If not, there are several excellent entomology reference books (see page 166) that provide dichotomous keys of the families in the major orders as part of their coverage.

A discussion of insect taxonomy would be nearly impossible without knowledge of the specialized jargon used by taxonomists. A section on Terminology has thus been provided to familiarize students with the vocabulary used in dichotomous keys, as well as in most field guides and reference books. The section on External Anatomy will discuss those insect features most often used in identification.

The remainder of this photo atlas and guide is devoted to covering the recognized insect orders and the major classes of arthropods. A discussion of the hierarchical system of classification and zoological nomenclature precedes the taxonomic coverage.

Dichotomous Keys

A taxonomic 'key' is a set of steps or clues, that if followed correctly should ultimately lead the user to the identification of an organism at a certain taxonomic level. For example, there are keys to the order level, keys to the family level, and so on all the way down to the species level. Keys have been developed to aid in the identification of most biological organisms, not just insects. Their use has become an integral component in most courses dealing with the life sciences such as biology, zoology, botany, and entomology. Therefore, the familiarity you gain with using dichotomous keys in entomology may well help you in other courses you plan to take.

The word 'dichotomous' means two-branching. It refers to the fact that the numbered steps of a key are presented in pairs or couplets. Each couplet presents two descriptions or sets of characters from which the user must choose that which most closely applies to the specimen under consideration. At the end of both of the descriptions in a couplet is a number (or a letter, depending upon what type of notation is being used). This number tells the user where to advance or which is the next couplet in the identification process. The user will always begin with couplet Number 1. However, the next couplet you are directed to may be 3, or 17, or 69, depending on the number of steps in the key. You will have achieved your identification when you reach a description that fits and provides you with a name at the end, rather than a new couplet number to which you should advance.

A simplistic three-step dichotomous key is provided below to illustrate certain important features (and pitfalls) of this common tool. Read through it and see if it makes sense to you. Try to identify problems in the wording of the descriptions and think of how you would modify it to improve its usefulness and make it less confusing and easier to use.

Sample Dichotomous Key

| 1A | Wings present. | 2 |
| 1B | Wings absent. | 3 |

| 2A | Wings covered with plumage. | **Bird** |
| 2B | Wings devoid of plumage. | **Bat** |

| 3A | Specimen large and gray with long trunk. | **Elephant** |
| 3B | Specimen small and brown with long tail. | **Monkey** |

Hopefully most college students know the difference between a bat, bird, elephant, and monkey, and can successfully identify them even without the use of the dichotomous key provided. However, the preceding exercise is useful in pointing out common problems with keys. For example, terms like "plumage" and "trunk" might not be familiar to everyone. Entomological keys often refer to specific anatomical features or use terminology that is only applicable to the particular insect group under consideration. It is therefore essential to learn the taxonomic terms and structures that make up the external anatomy of insects. The specialized vocabulary provided in this guide for the various insect groups discussed will make using a key treating a particular group less tedious and more fun.

Another thing to notice in the couplets of the preceding key are descriptions that use relative terms in the form of adjectives like "big" or "small" or "long". A good key will eliminate such non-specific adjectives and provide actual measurements. The measurement may be in units (i.e. the antennae are 10mm or more long) or stated as a comparison (i.e. the antennae are as long or longer than the body). The inclusion of a description of more than one character per couplet further increases the chances that the couplet will be interpreted correctly and a character recognized. This offers the student a fighting chance if the first character mentioned is unclear, or as is often the case, missing or broken on the particular insect specimen being examined. Legs, antennae, and wings on older fragile specimens break off very easily, sometimes leaving the student with nothing more than a desiccated body with which to deal.

When using a dichotomous key to make your determination, it's helpful to record the route you've taken in order to reach your goal. This allows you to backtrack in the unlikely event that you have made an error. With the help of your instructor, you should be able to pinpont exactly where it was that you went wrong. In the sample key, if your final identification was an elephant, you should have written: 1B, 3A Elephant. If you had determined that your specimen was a bird, you should have written: 1A, 2A Bird (unless you had a kiwi, which brings us to another problem).

You will learn quickly that keys don't always work. There are few absolutes in the world, but almost always exceptions. Murphy's Law dictates that the specimen you are trying to identify will be one of the exceptions that doesn't fit the descriptions provided in the key. Many different references and keys are available for the different insect groups and you are encouraged to use them. You are also encouraged to examine as many different specimens from the same group as possible. Insects are the most diverse creatures on Earth. Don't try to memorize just one or two in each family and hope that they will be the ones you will be tested on! Expand your knowledge and look at as many different specimens in your school's reference collection as possible. Remember, if you are permitted to use keys during your exams and quizzes, there is no reason why you should not be able to identify specimens of a given group, regardless of their origin. My Insect Taxonomy professor at Rutgers delighted in giving us both tropical and mimetic insects that we had never before seen in class. His reply to our complaints was to learn how to use the keys!

Often you need go no further than your own backyard to find an interesting example of insect fauna. This sphinx moth caterpillar was feeding on some weeds when the author spotted it in his garden. The linear shape of the larva combine with the eyespots to make it a convincing snake mimic. Many other caterpillars, especially in the swallowtail butterfly family, employ this type of defense.

External Anatomy

Introduction

When we look at insects today, we see an almost bewildering array of shapes, sizes, and colors. However, all insects have evolved from a common ancestral form and retain certain diagnostic features about their external anatomy. Scientists believe that the insect predecessor was elongate, roughly cylindrical, and segmented with paired appendages on each segment. Through time, certain segments grouped together into functional regions on the insect's body. Three such regions have resulted. Each region is called a **tagma** (plural = **tagmata**). They are represented on the insect body by the head, thorax, and abdomen.

The body wall of an insect is called its **exoskeleton**. It serves two functions. One is providing points of attachment for the muscles. The other is protection by means of the hard outer layer of the exoskeleton called the **cuticle**. This insect 'skin' is not one continuous hard shell, but rather is composed of a number of hardened plates called **sclerites**. These sclerites are separated from one another by seams and sutures, and by larger membranous areas such as between body segments. The degree of hardness depends upon how much **sclerotin** has been deposited. Extremely hard areas of the insect such as the jaws, or the wing covers in most beetles, are said to be heavily **sclerotized**. The more lightly sclerotized or membranous areas of the body permit the insect a degree of flexibility and greater range of movement.

The biology and behavior of a species will greatly affect its morphology, including whether it is hard- or soft-bodied. The exoskeleton of a scarab beetle can be incredibly hard. So hard that it may even cause a pin to bend when one is trying to mount it for a collection. Yet insects that live in protected situations such as internal parasites or the immature forms of social insects may be completely soft and vulnerable. Even the body of the scarab beetle is soft when it is an immature grub underground. Thus a great difference may exist in the same individual from one life stage to the next.

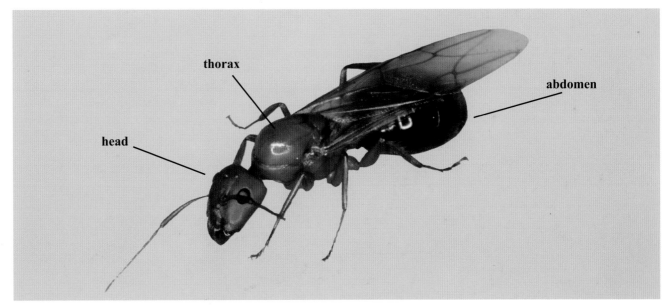

Fig. 1 The insect body is divided into three main regions or tagmata as seen on this queen ant. Formed by the fusion of body segments, they are the head, thorax, and abdomen.

Tagmata

Through the course of evolution the segments of insects have fused to form three body regions: the head, thorax, and abdomen. Each of these tagma have specialized external and internal structures that perform certain functions. The head is the main area of sensory perception and the point of ingestion. The thorax is located directly behind the head and contains the segments with the legs, and with the wings on adult insects. It is primarily responsible for locomotion. The abdomen is the hindmost tagma and follows the thorax. It contains the genitalia and point of egestion, as well as other specialized external structures. Internally, it contains many of the essential body systems.

Head

The head of most adult free-living insects is a hardened capsule. Internally, it contains the brain, musculature for the mandibles and mouthparts, and a girder-like system of braces called the **tentorium**. Its most noticeable external features are the eyes, antennae, and mouthparts. All of these are important characters used in the identification of insects. The basic areas of the head have been 'mapped out' and named. Learning the names below and their relation to head morphology will make it easier to describe and locate other structures. Examine a pinned or 'pickled' grasshopper specimen as you familiarize yourself with these terms.

Looking at the grasshopper face to face, you should see a line that comes straight down from the middle of the top of the head and then forks to give the impression of an inverted letter Y. This is the **epicranial suture**. The central area of the head or face occurring between the two branches of this suture is called the **frons**. The top of the head, from which the epicranial suture originates, is called the **vertex**. The side areas of the head (which we could think of anthropomorphically as the 'cheeks') are each called a **gena** (plural=**genae**).

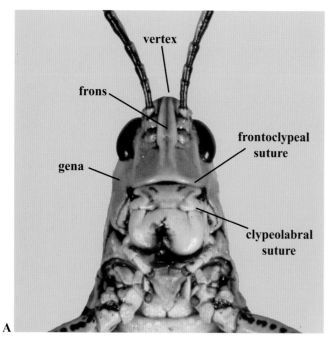 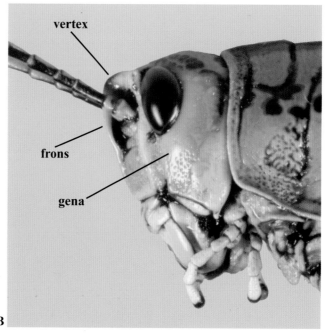

A　　　　　　　　　　　　　　　　　　　　B

Fig. 2　Frontal and lateral views of the head of a lubber grasshopper showing the basic features of the head and face.

Eyes

Insects have evolved various mechanisms for processing the light information from the environment in which they live. Vision may range from being extremely acute (as in many predators), to totally lacking (as in species that spend the majority of their life in the absence of light). One of the most prominent features of the insect head is a pair of multi-faceted **compound eyes**. In predators such as mantids or highly agile flying insects such as dragonflies and horse flies, the compound eyes may make up the majority of the head. The compound eye itself is made up of facets which may number from only a few to as many as several thousand. Each facet or separate visual unit is called an **ommatidium** (plural=**ommatidia**). The shape of the compound eyes, their location on the head, and whether or not they touch are all characters sometimes used in identification.

There may also be from zero to three simple eyes found on the head of adult insects. These **ocelli** (singular=**ocellus**) are composed merely of a single facet and can only detect changes in light intensity. Externally, they resemble small jewels embedded in the surface of the insect head, often amber in color. The term **stemmata** or **lateral ocelli** has been given to the simple eyes or photoreceptors found on the head of certain larval insects, as well as on the more primitive adults.

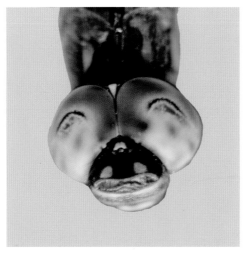

Fig. 3 Dragonflies are among the most agile fliers of the insect world. Much of the head is composed of the compound eyes.

Fig. 4 Like other insects that depend greatly on vision, horse flies have large compound eyes with many facets.

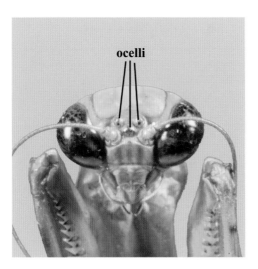

Fig. 5 Three small round ocelli are found on the head of this praying mantis just above and between the bases of the antennae.

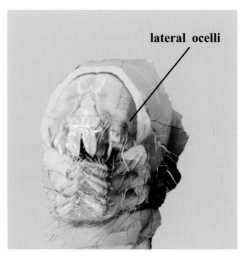

Fig. 6 A close examination of the head of a caterpillar may reveal the light sensitive stemmata or lateral ocelli.

Antennae

The antennae or 'feelers' are often the most noticeable appendages on the insect head (and the structures most prone to breaking on pinned specimens!). There is a great variety of shapes and sizes from the long thread-like antennae of katydids that may be twice the body length, to the tiny bristle-like antennae of dragonflies that could be easily overlooked. The type of antennae is one of the key taxonomic features that enables us to identify insects down to the family level. Anatomically, the antenna has been divided into three sections. The basal antennal segment is called the **scape**. The second antennal segment from the base is the **pedicel**. All the rest of the antennal segments, whether there be one or fifty, are referred to collectively as the **flagellum**. Each individual segment on the flagellum is called a **flagellomere**.

Types of Antennae

Filiform - Thread-like or hair-like. Composed of a series of cylindrical or flattened segments. Examples include cockroaches, crickets, katydids, grasshoppers, true bugs, and bark lice. (Fig. 7)

Moniliform - Bead-like and composed of a series of rounded segments like on a pearl necklace. Examples include termites and some beetles. (Fig. 8)

Geniculate - Elbowed. Abruptly bent, such as in a knee joint or elbow joint. Examples include ants, bees, and weevils. (Fig. 9)

Serrate - Combination of roughly triangular segments that give a saw-toothed appearance. They are found on click beetles and others. (Fig. 10)

Setaceous - Slender, bristle-like, and gradually tapering to a tip. Examples include dragonflies, damselflies, and cicadas. (Fig. 11)

Plumose - Feather-like, or with whorls or clumps of hairs. Giant silk moths and mosquitoes both have plumose antennae, which are more pronounced on the males of each. (Fig. 12)

Pectinate - Comb-like. Lateral processes stick out from the antennae at regular intervals like the teeth of a comb. Examples include glow-worms and some fireflies. (Fig. 13)

Capitate - The tip of the antenna is enlarged into a rounded knob. Examples are found on butterflies, antlions, and owlflies. (Fig. 14)

Clavate - The tip of the antenna is enlarged into a broadened club. Carrion beetles and carpet beetles both have distinctly clubbed antennae. (Fig. 15)

Lamellate - Clubbed antenna where the terminal segments are enlarged parallel plates that stick out perpendicular to the rest of the antenna. Scarab beetles typically have lamellate antennae. (Fig. 16)

Aristate - Three-segmented antenna where the third segment bears a protruding hair called an **arista**. Found on blow flies, house flies, flesh flies, and tachinid flies as well as others. (Fig. 17)

Stylate - Antenna terminates in a long slender point called a **style**. The style may be similar to an arista, but occurs on the tip and not on the side. Examples include robber and bee flies. (Fig. 18)

Types of Antennae

Fig. 7   Filiform or thread-like antennae of an earwig.   Order Dermaptera

Fig. 8   Moniliform or bead-like antennae of a termite.   Order Isoptera

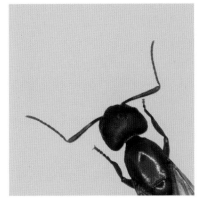

Fig. 9   Geniculate or elbowed antennae of an ant.   Order Hymenoptera

Fig 10   Serrate or sawtoothed antennae of a lycid beetle.   Order Coleoptera

Fig. 11   Setaceous or bristle-like antennae of a cicada.   Order Homoptera

Fig. 12   Plumose or feathery antennae of a giant silk moth.   Order Lepidoptera

Fig. 13   Pectinate or comb-like antennae of a firefly.   Order Coleoptera

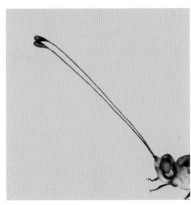

Fig. 14   Capitate or knobbed antennae of an owlfly.   Order Neuroptera

Fig. 15   Clavate or clubbed antennae of a carrion beetle.   Order Coleoptera

Fig. 16   Lamellate antennae of a scarab beetle.   Order Coleoptera

arista

Fig. 17   Aristate antennae of a tachinid fly.   Order Diptera

Fig. 18   Stylate antennae of a bee fly.   Order Diptera

Mouthparts

The remaining significant features of the head are the mouthparts. In some species that do not feed as adults, the mothparts may be **vestigial** or absent. However, most adult insects will have distinctive mouthparts whose shape and morphology are indicative of the type of food it consumes. We shall discuss the basic mouthparts as they occur in a chewing mouthtype such as is found in a cockroach or grasshopper. While this will familiarize us with basic oral anatomy and its terminology, the examination and discussion of other mouthtypes will show the extreme modifications that have evolved from this more primitive chewing form.

Chewing Mouthtype

If we look at a grasshopper head-on ('frons to frons'), we should be able to see some of the insect's mouthparts. How many of the mouth structures that are easily visible will depend on their position at the time the insect was preserved. However, changing the orientation of a pinned specimen for a different view, or the removal and dissection of parts on alcohol specimens, will permit the observation and examination of all the major mouthparts. At first glance, you will probably notice a couple of flap-like things in the front, perhaps a hard brown to black thing on each side, and one to two dangly jointed things on each side as well.

The roughly rectangular flaps that occur on the bottom of the grasshopper face, one over the other, are the **clypeus** and **labrum**. The immovable clypeus is higher up, occurring just below the frons and separated from it by a suture (frontoclypeal suture). The labrum can be considered the 'upper lip' of chewing insects. It occurs just below the clypeus and can be bent up along the seam where the two meet. Deeper into the face on a level below the clypeus (and sometimes obscured by it) are the **mandibles**. The mandibles are paired structures and occur on both sides of the mouth, meeting in the middle. They are responsible for tearing or biting off chunks of food and grinding it or chewing it to a consistency that the insect can ingest. The darkest portions of the mandibles are the hardest or most heavily sclerotized. Looking at a mandible under the microscope will show areas that are sharp for cutting and others that are blunt for grinding. These areas are comparable to our incisors and molars.

Just below each mandible is a **maxilla** (plural=**maxillae**). Each maxilla has both a fleshy portion and a sharp hardened portion that is also used for cutting. Projecting from near the base of each maxilla is a five-segmented appendage that almost resembles a small leg. This is the **maxillary palpus** (plural=**palpi**) and serves to help manipulate the food to the mouth, as well as sensory functions. Observations of the movements of palpi on a live cockroach or grasshopper that is feeding will show them to be quite dextrous. Below the maxillae is a single structure that can be thought of as the 'lower lip' and is called the **labium**. A three-segmented **labial palpus**, which is similar in form and function to the maxillary palpus (only slightly smaller), projects from the base of each side of the labium. The tongue or **hypopharynx** in insects with a chewing mouthtype is a fleshy knob on the upper surface of the labium. The actual mouth occurs between the labrum and the hypopharynx.

Chewing Mouthparts

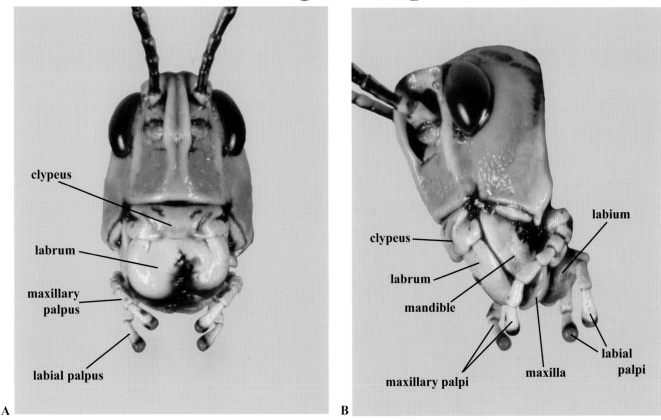

clypeus

labrum

maxillary palpus

labial palpus

A

clypeus

labrum

mandible

maxillary palpi

labium

maxilla

labial palpi

B

Fig. 19   Frontal and lateral view of the head of a lubber grasshopper showing the structures that compose the mouthparts and their appendages.

Clypeus and Labrum

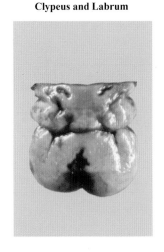

Right Maxilla

Right Mandible

Left Mandible

Left Maxilla

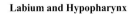

Labium and Hypopharynx

Fig. 20   The anatomical structures that form the chewing mouthparts of a lubber grasshopper presented in a disarticulated state.

Chewing Mouthparts

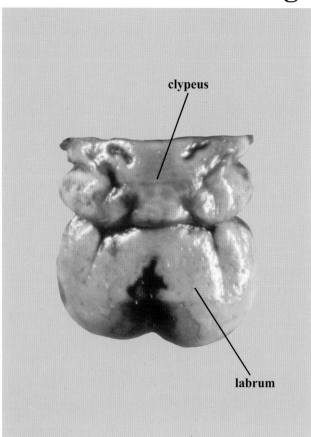

Fig. 21   External surface of the clypeus and labrum of a lubber grasshopper.

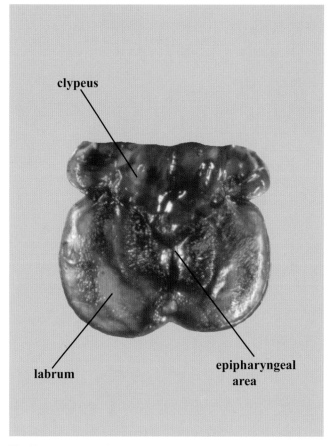

Fig. 22   Internal surface of the clypeus and labrum of a lubber grasshopper.

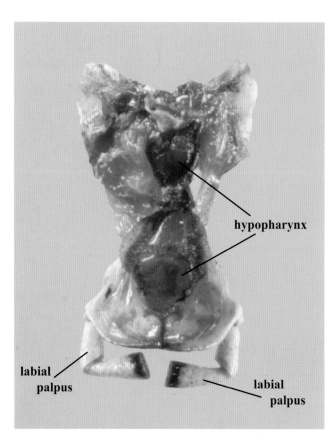

Fig. 23   Internal surface of the labium of a lubber grasshopper showing the tongue (hypopharynx) in place.

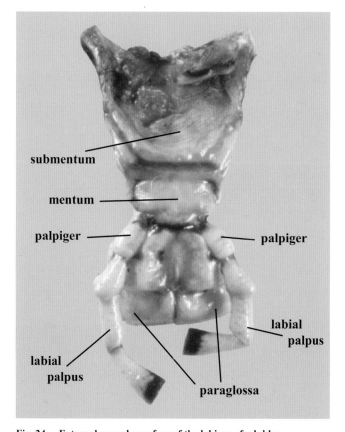

Fig. 24   External or undersurface of the labium of a lubber grasshopper.

Chewing Mouthparts

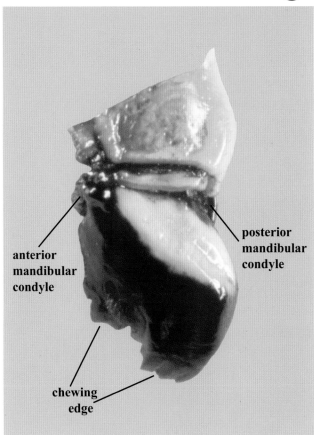

Fig. 25   Left mandible of a lubber grasshopper. Note that a considerable portion of the grasshopper head where the mandible attaches is visible in this photo.

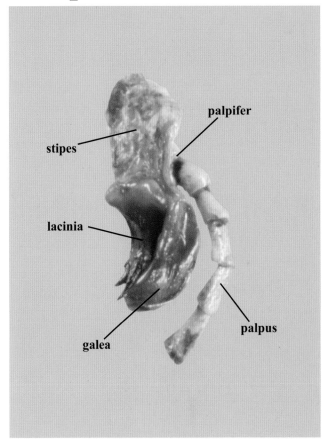

Fig. 26   Left maxilla of a lubber grasshopper.

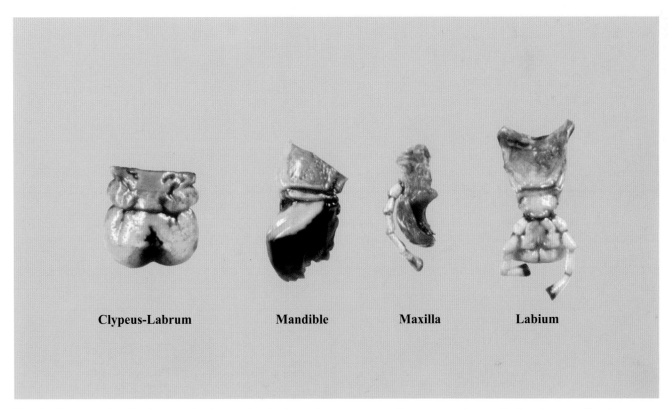

Fig. 27   Representative disarticulated chewing mouthparts of the lubber grasshopper aligned to provide an indication of comparative sizes.

Piercing-Sucking Mouthtype

Many insects feed on liquid rather than solid food. The nourishing liquid may take the form of nectar from flowers, or plant juices withdrawn directly from plugging into the plant's vascular system, or even blood that is sucked out of an animal's vascular system. A piercing-sucking mouthtype is responsible for the removal of a liquid food source and the mouthparts themselves are generally referred to as a beak. These beaks can be formidable, as evidenced by the predatory assassin bugs that feed on other insects. Beaks may also be slender and stiletto-like, as we find associated with a cicada. An intermediate-sized beak that most of us are all too familiar with is found on most mosquitoes.

The piercing-sucking beak is composed of the same basic mouthparts found in chewing insects, but adapted for piercing tissue and removing liquid. The mandibles and maxillae are usually long, sharp **stylets** housed within a protective sheath often derived from the labium. One or more channels are formed by the arrangement of the mouthparts. A food channel is used to convey the liquid from its source to the insect's mouth. A salivary channel serves to introduce secretions from the insect into the food source. These may be toxins that stun or kill their prey, enzymes that break down and digest tissue, or anti-coagulents such as those used by mosquitoes and other blood feeders.

Sponging Mouthtype

Various species of flies such as the common house fly and blow flies exhibit sponging mouthtypes. Instead of a piercing beak, the structures have evolved into a mechanism that permits the non-invasive sucking of liquids. The proboscis is composed of a basal portion called the **rostrum**. A pair of forward protruding maxillary palpi originate from this part. Between the rostrum and the tip of the mouthparts is the **haustellum**. At the end of the haustellum are two fleshy lobes called **labella** (singular=**labellum**). Each labellum is crossed with transverse grooves which by means of capillary action draw liquid in and up to the food channel. The secretion of saliva onto the food medium helps to break it down and liquefy it so that it can be sucked up.

Variations on the pure sucking mouthtype described above are found in a number of fly species that feed on blood. For example, the stable fly and the tsetse fly have teeth on the labella that cut into the skin to obtain blood. Their labella lack the grooves and the blood is taken in through the physical action of a sucking pump in the insect's head. Other blood-feeding flies like the horse fly and the deer fly have retained the larger grooved labella, but evolved piercing mandibles and maxillae. They pierce the skin with these sharp stylets and then sponge up the pooling blood that is freed as a result of it. Such a mouthtype might be considered a piercing-sponging type.

Siphoning Mouthtype

Most butterflies and moths have very obvious mouthparts. There is a tongue that has the form of a coiled tube. In addition, there is usually a pair of hairy or 'furry' looking structures that stick almost straight out, perpendicularly from the head. These hairy things that project from the base of the head are the labial palpi. The coiled tongue is composed of parts called the **galeae**, which are derived from the maxillae. The other mouthparts are either too small to see easily or have been 'lost' evolutionarily. The tongue or proboscis of a butterfly or moth functions much like a straw. It is inserted into a liquid such as flower nectar, which is sucked up via a food canal in its center. The two halves of the tongue are kept together by physical structures that interlock.

Non-Chewing Mouthtypes

Fig. 28   Piercing-sucking mouthparts of an assassin bug. Family Reduviidae

Fig. 29   The piercing-sucking beak or proboscis of a mosquito in the genus *Psorophora*. Family Culicidae

maxillary palpi

labellum

Fig. 30   Sponging mouthparts of a blow fly. Family Calliphoridae

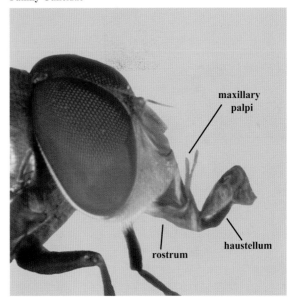

maxillary palpi

rostrum   haustellum

Fig. 31   Sponging mouthparts of a blow fly. Family Calliphoridae.

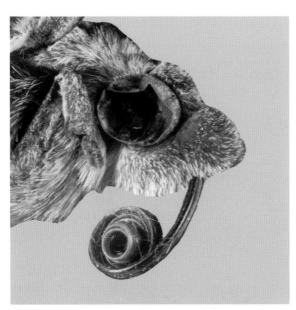

Fig. 32   The coiled tongue of a sphinx moth illustrates the siphoning mouthtype found in the order Lepidoptera. Family Sphingidae

Fig. 33   Only the butterfly genus *Heliconius* feeds on pollen which is gathered on the tongue.   Family Heliconidae

Chewing-Lapping Mouthtype

Some insects have evolved to feed on liquids, but while still retaining their chewing mandibles. This situation is seen in certain species of bees. For example, the honey bee (Fig. 34) does not use its mandibles for feeding, yet needs them for molding wax to make combs and for chewing its way out of its pupal cell. The nectar it gathers and feeds on is obtained through a retractable tongue made up of parts of the maxillae and labium. A sucking pump helps convey the nectar through the tongue to the mouth.

Additional Mouthtypes

There are probably variations on all the major mouthtypes that we have discussed, each of which have evolved in response to specialized life styles and feeding situations. The immature or larval stages of insects often possess mouthtypes that differ greatly from those of the subsequent adult. Butterflies and moths demonstrate this fact well. Most caterpillars have chewing mouthtypes, while most adult butterflies and moths have siphoning mouthtypes. An examination of the immature forms of selected insects will reveal some interesting adaptations.

The immature stages of dragonflies and damselflies live in water and are called larvae or **naiads**. Body structures used in breathing and locomotion have been modified due to their aquatic environment. Unique to this order (the Odonata) of insects however, is its mouthtype and how it is used in prey capture. The developing dragonflies/damselflies feed on small water creatures including tadpoles and even fish. The labium (Fig. 35) has become a long, flattened hinged structure, up to one-third the length of the body. It doubles back on itself and is concealed when not actively being used. The labial palpi have become thickened and armed towards the tips with movable hooks. When prey swim by the immobile naiad, the hinged labium shoots forward and either impales the item on the hooks or grabs the victim and draws it to the mandibles.

The larval stages of lacewings and antlions share a similar mouthtype, although their method of prey capture is very different. Both groups have long, sickle-like jaws that are grooved. This groove or **sulcus** acts as a channel to direct the body juices of the prey to the mouth of the predator. Such a mouthtype can be called **sulcate**. Lacewing larvae are active hunters with smooth curving mandibles (Fig. 36), and often feed on aphids. Antlion larvae are 'sit and wait' predators that bury themselves at the bottom of tiny pits. If a hapless ant falls in the pit, a pair of long, straight, toothed jaws (Fig. 37) emerge to impale it when it comes to rest on the bottom.

The last mouthtype we shall discuss belongs to a larval fly or maggot. Using a blow fly as an example, it is easy to see two darkened, heavily sclerotized **mouth hooks** (Figs. 38-39) that contrast with the light-colored body of the maggot. The head is found at the narrow end of the larva, while the **terminal spiracles** or respiratory openings are found at the broad posterior end. Maggots are voracious feeders and in large numbers can completely remove the flesh from a carcass in a matter of a few days.

Non-Chewing Mouthtypes

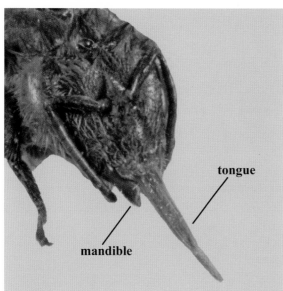

Fig. 34   Chewing-lapping mouthparts of a bee with the tongue extended.   Family Apidae

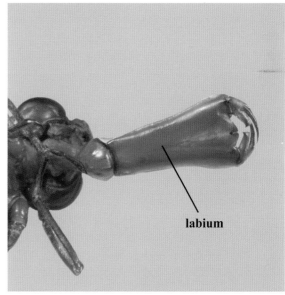

Fig. 35   Extended labium of a dragonfly larva viewed from beneath showing the terminal spines that are used to impale their prey.   Order Odonata

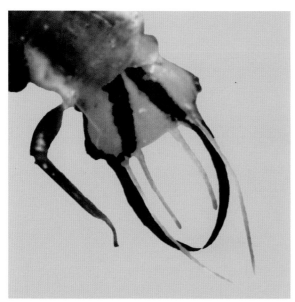

Fig. 36   Large curving mandibles are the primary structures of the sulcate mouthparts seen on this green lacewing larva.
Family Chrysopidae   *Courtesy ARS/USDA*

Fig. 37   The sulcate mouthparts of the antlion are formed mainly by the huge toothed grooved mandibles.
Family Myrmeleontidae

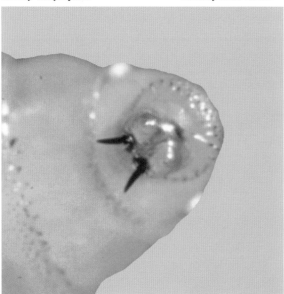

Fig. 38   The paired darkened mouth hooks of this carrion fly larva or maggot are used for tearing flesh and ripping tissue.
Family Calliphoridae   *Courtesy ARS/USDA*

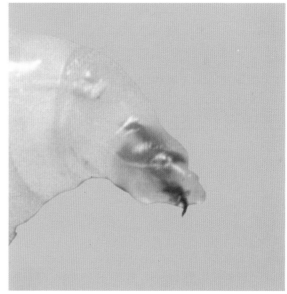

Fig. 39   In lateral view the black pointed mouth hook projects below the body while the darkened cephalo-pharyngeal skeleton is seen through the body wall.

Thorax

Segmentation

Working our way back from the anterior end of the insect, we encounter a tagma directly behind the head composed of three fused segments and called the thorax. The obvious external structures (legs and wings) found associated with the thorax of most adult insects make it clear that it is the center of locomotion.

The individual segments making up the thorax, from anterior to posterior, have been named the **prothorax**, **mesothorax**, and **metathorax**. The prefixes pro-, meso-, and meta- are commonly used when discussing thoracic structures in order to more specifically identify upon which segment they are located. This convention is often used in dichotomous keys to insect taxa. Therefore, a spiracle found on the first thoracic segment or prothorax would correctly be referred to as a prothoracic spiracle. A leg on the second thoracic segment would be called a mesothoracic leg. A wing on the third and last thoracic segment would be a metathoracic wing.

The thoracic region is usually heavily sclerotized to provide support and bracing for the movement of the legs and wings. The top portion of each thoracic segment is called a **notum** (plural=**nota**). The pronotum is often prominent and conspicuous. The bottom portion of each thoracic segment is known as a **sternum** (plural=**sterna**). Each side area of the body located between the notum and the sternum is called a **pleuron** (plural=**pleura**).

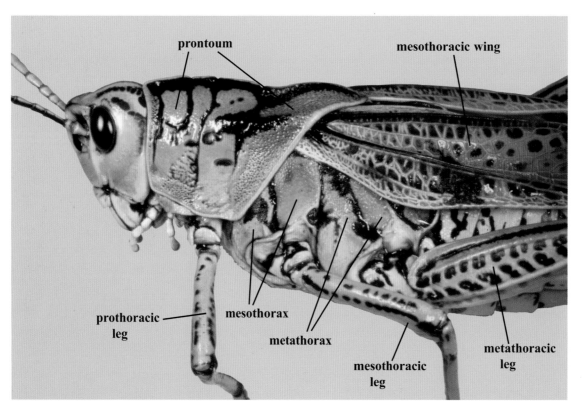

Fig. 40 Lateral view of the thorax of a lubber grasshoper showing the appendages and features associated with this tagma.

Legs

Segmentation

Most adult insects have six legs, which is why the Class Insecta was formerly called the Class Hexapoda. One pair of legs originates from each thoracic segment. As with the rest of of the insect body, knowledge of the parts and anatomy of the legs is essential when discussing the similarities and differences between taxonomic groups. The typical adult insect leg has five major parts or segments. The first segment joining the leg to the body is the **coxa** (plural=**coxae**). Progressing outward, the next segment is the **trochanter**. (In many insect groups, the coxae and trochanters are small squarish or collar-like segments.) Following the trochanter is the **femur** (plural=**femora**). After the femur is the **tibia** (plural=**tibiae**). The femora and tibiae are usually the longest and most evident portions of the insect leg. The terminal portion of the leg is the **tarsus** (plural=**tarsi**), which may be composed of from one to five segments or **tarsomeres**. The last tarsal segment often has a tarsal claw.

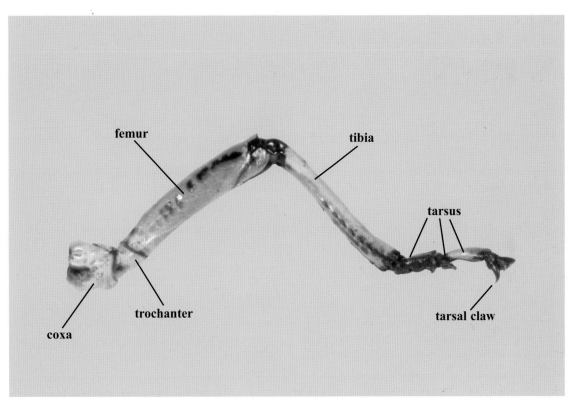

Fig. 41   Mesothoracic leg of a lubber grasshopper with anatomical features labeled.

Leg Types

An insect living underground that is forced to contend with moving through a heavy medium such as soil or sand will show morphological adaptations extremely different from species that run on the soil surface. Insects living in an aquatic environment will exhibit anatomical structures that facilitate living in or under water. While the insect body as a whole will reflect modifications for living within a certain physical habitat, some of the most severe and noticeable morphological changes have occurred on the legs.

Raptorial

Just as the word raptor gives rise to the image of a bird of prey (or predatory dinosaur), so should the term **raptorial** make one think of a predatory insect. Raptorial legs are characterized by sharp teeth or spines that impale and cling to prey organisms. The praying mantis (Fig. 42) is probably the most commonly recognized insect with raptorial legs, but variations on this leg type are seen throughout the insect world. Many hunting species have evolved legs for grasping prey, including assassin bugs, ambush bugs, diving beetles, dragonflies, mantidflies, and even rare predatory caterpillars!

Fossorial

Life underground or beneath the leaf litter is termed **fossorial**, as is the leg type evolved for locomotion and movement underground. The front or prothoracic legs are usually the most affected, and sometimes the only pair that have become modified. Shovelling through the soil requires a wide, flattened, blade-like apparatus. This is exactly what we see in the heavily sclerotized front legs and claws of the mole cricket (Fig. 43). Their powerful digging ability and torpedo-like body shape permit them to move with relative ease in galleries they excavate beneath the earth. The underground immature stages of cicadas whose 'shells' can often be found on tree trunks in the summer, also show this adaptation.

Cursorial

Long slender legs used for running are classified as **cursorial**. Examples are cockroaches (Fig. 44), crickets, and many of the predaceous beetles such as the tiger and ground beetles. These legs are adapted for quick movement over a variety of surfaces. Anyone who has tried to catch or corner a cockroach knows how quickly they can scurry away.

Saltatorial

Those who speak Spanish will recognize a portion of the word *saltar*, which means 'to jump'. The enlarged, muscular-appearing hind leg (metathoracic leg) of a grasshopper (Fig. 45) is a good example as a **saltatorial** leg. The hind femora are much bigger than the others and can be used to launch the grasshopper into the air. Once airborne, it can take flight or complete its leap. Coupled with the cryptic coloration of many grasshoppers, jumping is an effective get-away mechanism. The katydids and some crickets also have saltatorial rear legs.

Natatorial

Many insects that live in aquatic environments have evolved changes in leg morphology that permit them to effectively propel themselves through the water. Such adaptations are evident on the water boatmen and backswimmers among the true bugs, and on the water scavenger and predaceous diving beetles (Fig. 46) as well. One pair of legs (usually the metathoracic) are extremely flattened and oar-like. The surface area of these leg segments is often increased through the presence of long, stiff hairs that fringe them. If examination of a pinned specimen reveals natatorial legs, it is almost certainly an aquatic species.

Scansorial

Many species exist in the insect world that are obligate **ectoparasites** of mammals. Arthropods such as fleas, lice, and ticks live in close association with their hosts, spending much of their life in the host's hair or fur. Specialized legs adapted for climbing among hair or fur have been named **scansorial**. If we look at the meso- and metathoracic tarsi of the human pubic or crab louse (Fig. 47), we will see large claws that have a gap of particular size between them. This gap closely matches the diameter of a human pubic hair which these lice use as supports as they move about or feed on their hosts.

Types of Legs

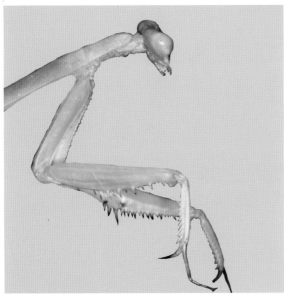

Fig. 42   The raptorial front legs of the praying mantis are covered with spines and used for grasping prey. Family Mantidae

Fig. 43   The mole cricket has broad flattened clawed fossorial front legs that enable it to move beneath the soil. Family Gryllotalpidae

Fig. 44   The cursorial legs of the cockroach are long and slender enabling it to run at great speeds. Family Blattidae

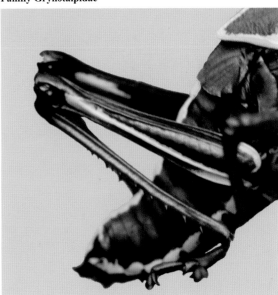

Fig. 45   The enlarged femur on this grasshopper hind leg is used for jumping and termed saltatorial. Family Acrididae

Fig. 46   Natatorial legs are used for swimming and typically are flattened with long hairs like in this predaceous diving beetle.   Family Dytiscidae

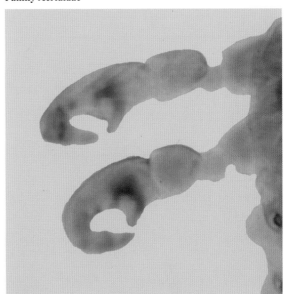

Fig. 47   Scansorial legs, such as on this crab louse, are used for clinging.   Family Phthiridae
Photo JL Castner   Courtesy IFAS/UF

Wings

The final structures associated with the thorax that we shall discuss are the wings. An important feature in the success of insects as a group, functional wings are almost always found only on adult insects. However, not all insects have wings. Some of the most primitive groups are **apterous** or wingless. Other more advanced groups of insects have 'lost' their wings evolutionarily as a survival adaptation to a specialized way of life. For example, both the fleas and the sucking lice are apterous, probably because wings would be a hindrance to their movements among the fur and between the hairs of their hosts.

Most winged insects have two pairs of wings. One pair each is located on the mesothorax and metathorax, which are collectively called the **pterothorax**. Some only have one pair of wings (flies), while in others winged individuals represent the reproductive form and are produced only at certain times of the year (ants and termites). Other groups, as indicated above, have no wings. In our continuing effort to become familiar with all parts of the insect anatomy, we shall discuss the common features and terminology associated with the generalized insect wing. Structures or modifications found only in a specific group of insects will be covered under the section treating that group.

Take a pinned fly or honey bee with the wings spread perpendicular to the body and view it from above. Arrange the body of the insect with the head in the twelve o'clock position and the tail in the six o'clock position. If we first look at just the outer edge of the wing, we can divide it into three regions (Fig. 48). The front or 'leading' edge is called the **costal margin**. The edge encompassing the wingtip is the **apical margin**. The rear edge is termed the **anal margin**. By convention, wings are usually illustrated in the position described above unless otherwise stated.

Dark thickened 'lines' appear to radiate out and crisscross throughout the wing (Fig. 49). These are wing veins and are very important characters for identification in some groups. In other groups, venation is greatly reduced or almost non-existent. The veins connect to the insect body and circulatory system. The movement of blood through these veins is essential in expanding the wings to their full size after the insect has shed its skin for the last time. The long veins originating near the base of the wing and more or less parallel to the costal margin are called **longitudinal veins**. The names of the main longitudinal veins from costal edge to anal edge are the **costa**, **subcosta**, **radius**, **media**, **cubitus**, and **anal** veins. The short veins that connect one longitudinal vein to another are called **cross veins**. Between the various veins are clear or membranous portions of the wing termed **cells**. If the cell has any part of its perimeter on the wing edge, it is called an **open cell**. If the cell is entirely closed by longitudinal and cross veins however, it is called a **closed cell**.

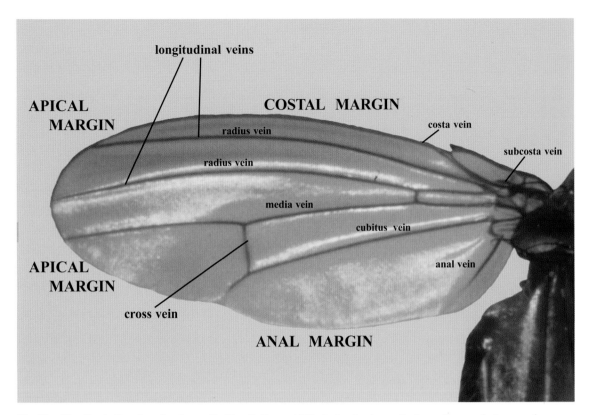

Fig. 48 Venation in the wing of a vinegar fly (Family Drosophilidae) showing types of veins and areas of wing margins commonly used in taxonomic keys. Note that the radius vein splits near the base of the wing, which is why there appears to be two with the same name.

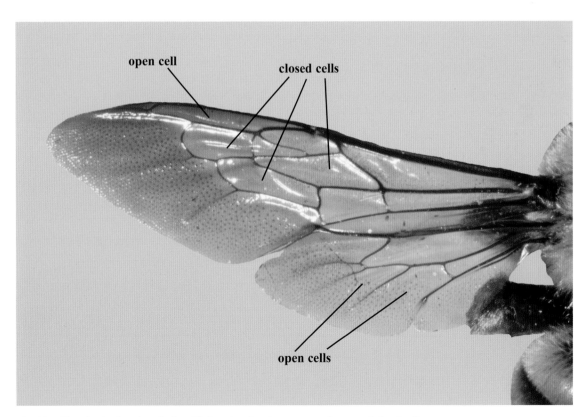

Fig. 49 Venation in the wing of a bumble bee showing the presence of open and closed cells.

Abdomen

Segmentation

The abdomen is the third and final tagma, and forms the posterior portion of the insect's body. It contains organs whose functions deal mainly with reproduction, digestion, circulation, and respiration. Only the external structures will be treated here. There are typically eleven abdominal segments although a reduction in number has occurred in some groups. Each segment appears to be divided into a top and bottom part. The top sclerite is called the **tergum** (plural=**terga**). The bottom sclerite is called the **sternum** (plural=**sterna**). There are no pleural sclerites as in the thorax, but the membranous area between the terga and the sterna is called the **pleural membrane**. Coupled with the membranous intersegmental areas, it permits flexibility and expansion or extension of the body.

External Structures

One of the most obvious abdominal features in most insects occurs at the tip of the abdomen. This is a pair of feeler-like structures called **cerci** (singular=**cercus**). The cerci vary in size among different groups of insects, but are usually easily visible. They appear to be sensory organs that are useful in detecting vibrations and disturbances in air currents. However, in some insects like the earwigs, they have evolved into sclerotized pincers that are used for both defense and prey capture. The cerci of silverfish are extremely long and tail-like, and are accompanied by a third similar structure between them called a caudal filament.

Another feature found at the tip of the abdomen, but only in female insects, is the **ovipositor**. This structure is used by the female when laying her eggs. It allows for the careful and exact placement of the eggs in some cases, and sometimes provides the mechanical means for introducing the eggs into a specific material such as soil, wood, or a plant stem. The ovipositor may be large and obvious as in some katydids, or it may be withdrawn and hidden in the body as in the female house fly. The latter type has segments that neatly 'telescope' one within the other. The ovipositors of some ichneumon wasps are extremely long and are used to drill through the solid wood of logs to reach wood-boring beetle larvae that live within. Other wasps and bees have ovipositors that are modified for defense that some of us are all too familiar with. These ovipositors are called stingers and can be extruded out to jab intruders, often with the injection of venom as well. Female honey bees, bumble bees, yellowjacket wasps, paper wasps, and hornets protect themselves and their nests in this way.

Copulatory structures are also present at the tip of the abdomen. In male insects, the penis or **aedeagus** is often withdrawn into the body. Accessory structures such as **claspers**, used to hold the female during mating, are also sometimes present on males. In most insects, sperm transfer is direct from the male to the female during copulation or mating.

A series of small oval spots, one pair per segment, may also be present along each side of the abdomen. They are easily visible on most caterpillars and are called **spiracles**. They are the external openings of the insect's respiratory system. They connect to chitin-lined tracheal tubes that get smaller and smaller as they ramify throughout the insect's body. By coordinating muscular movement with the opening and closing of the spiracles, the insect can effectively ventilate air through its body.

Abdominal Structures

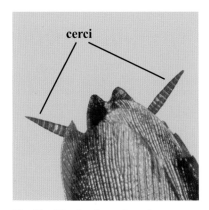

Fig. 50   The tip of a cockroach abdomen shows large, segmented cerci. Family Blattidae

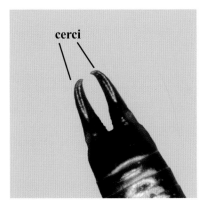

Fig. 51   At the tip of the earwig abdomen are forceps-like cerci that are used in defense and prey capture. Order Dermaptera

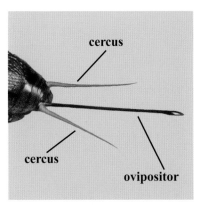

Fig. 52   On the abdomen of the female field cricket are two long slender cerci and a single central needle-like ovipositor. Family Gryllidae

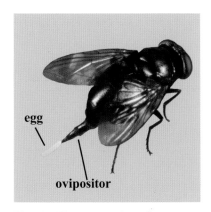

Fig. 53   The narrow telescoping terminal segments of some fly abdomens serve as an ovipositor. Here an egg has just been laid. Family Calliphoridae

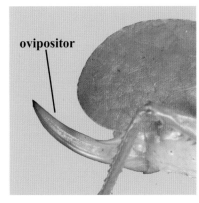

Fig. 54   Katydids lay their eggs in a wide variety of materials including wood. A sword-like ovipositor like this one is used to drill through bark. Family Tettigoniidae

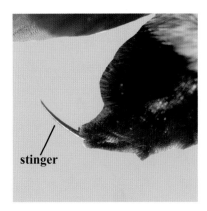

Fig. 55   Some ovipositors have been modified into weapons and are usually called stingers. The stinger of this bumble bee is extruded from the tip of the abdomen.   Family Apidae

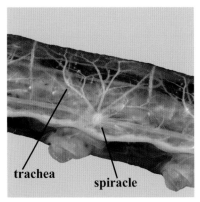

Fig. 56   Silvery cuticle-lined tracheae branch out from the spiracle of this skipper larva which can be seen as a white oval above the caterpillar's leg. Family Hesperiidae

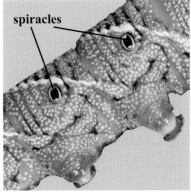

Fig. 57   The spiracles of this sphinx moth caterpillar are seen as dark oval structures above the legs high up on the sides of the body. Family Sphingidae

Fig. 58   The paired posterior spiracles of some fly larvae are darkened and easily visible as in this flesh fly maggot. Family Sarcophagidae

Insect Development

Introduction

Insects pass through a series of stages when passing from egg to adult. The appearance of these stages and the time spent within each varies with the species and with the environmental conditions that are present. For example, the developmental cycle is usually accelerated as temperatures increase. The process of undergoing physical changes from one life stage to the next is known as **metamorphosis**. This is accomplished by means of the insect 'shedding its skin' or undergoing **ecdysis**. The act of shedding the skin is also known as **molting** (Fig. 59). This occurs at certain points as the insect grows. The old shed skin that is left behind is called an **exuvium** (Fig. 60). The time period spent in any particular life stage is referred to as a **stadium**, while the insect itself may be referred to as an **instar**, especially during larval development. For example, a second larval instar is an insect that has shed its skin once to proceed to its second stage since hatching from an egg.

Ametabolous Development

The simplest form of development occurs when the only differences between the immature and adult forms are size and sexual maturity. Very primitive insects such as silverfish and springtails (Fig. 61) show this development, which is called **ametabolous**. The young insect hatches from an egg and is essentially a small version of the adult. There is no outward morphological difference between immatures and adults except size. The insects that undergo ametabolous development are all wingless.

Hemimetabolous Development

A slightly more complex form of growth is **hemimetabolous development** or **gradual metamorphosis**. Eggs are laid and hatch out into immature forms that are called nymphs. These nymphs are generally similar to the adults, but have no wings. With each shedding of the skin, the nymph becomes a little bigger. After ecdysis occurs several times, small wing pads begin to appear on the outside of the nymph. These pads get larger after the skin is shed each time until the adult stage is reached. At this point, the wings have developed to full size. Since the wings develop externally on the body, these insects are also described as **exopterygotes**. Among insects whose immature and adult stages are terrestrial, hemimetabolous development is seen in true bugs (Fig. 62), crickets and grasshoppers (Fig. 63), mantids, cicadas, cockroaches (Fig. 64), and others

Within the insect groups that undergo hemimetabolous development are several orders whose immature stages are aquatic. These are found in the mayflies, stoneflies (Fig. 65), and dragonflies and damselflies (Fig. 66). The aquatic young possess gill structures that permit them to breathe in water, and although referred to here as nymphs, they may be called either **naiads** or larvae in other texts. The morphology of these aquatic immatures differs considerably from the adults due to the differences in larval and adult habitats and ecology. When ready to molt to the adult stage, the fully adult nymph will either come to the water's surface or exit the water completely to shed its skin. In older works, insects that develop in this manner were the only ones described as undergoing hemimetabolous development. The older term for the development of terrestrial insect orders with nymphs was **paurometabolous development**.

Insect Development

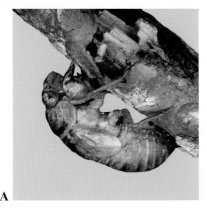

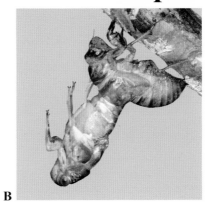

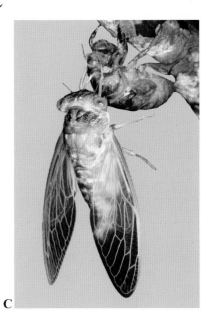

Fig. 59 The metamorphosis of a cicada from nymph to adult. (A) A mature nymph has climbed up to a branch to molt after having spent years underground. (B) The nymphal shell or skin has split and the new adult has almost pulled itself all the way out. (C) In a matter of minutes, the tiny wrinkled wings extend to their full size. The adult hangs on the shell until it hardens. (D) As the tissues harden or sclerotize, the colors darken and the adult takes on the appearance it will have for the rest of its life. After several hours, it is ready to fly and seek a mate.

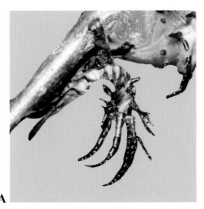

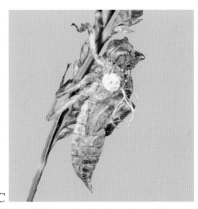

Fig. 60 The shed skins or exuvia of various insects. (A) The old larval skin of the hickory horned devil caterpillar retains the large frontal horns that form such a visible part of the caterpillar. (B) The nymphal shell of a cicada, found on a tree trunk where it emerged to an adult. (C) The larvae of dragonflies climb up on aquatic vegetation to molt to an adult. Their shed skins are often left behind on the plants.

Fig. 61 Very primitive insect groups undergo ametabolous development where there is little difference in the young and the adults, except for their size. Two such groups are: (A) silverfish which belong to the Order Thysanura and (B) springtails which belong to the Order Collembola.

27

Gradual Metamorphosis

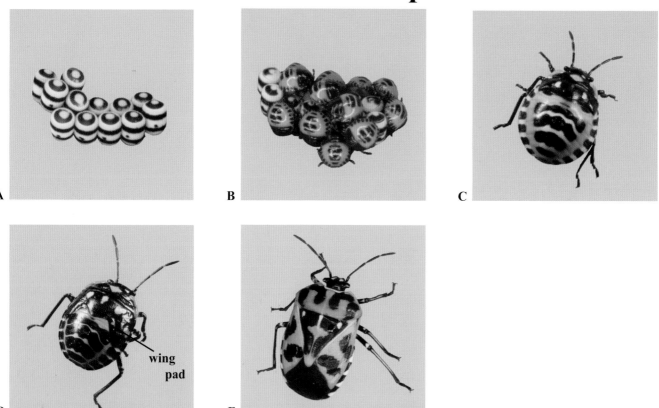

Fig. 62   The life cycle of the harlequing stink bug (Family Pentatomidae). (A) The barrel-like eggs are laid in small clusters on the foodplant. (B) The first instar nymphs remain close to the eggs shortly after hatching. (C) This third instar nymph retains the same shape and colors as the earlier stages, but has increased considerably in size. (D) Small dark wing pads are visible on this mature nymph. (E) The adult has full-sized wings that cover the entire dorsum of the body.

Fig. 63   The life cycle of the American grasshopper (Family Acrididae). (A) A clutch of grasshopper eggs is deposited to a depth of several inches in the ground. (B) The first instar nymphs hatch from the eggs and make their way to the soil surface. (C) Second instar nymphs show some slight changes in color. (D) In later nymphs, tiny wing pads become visible just behind the pronotum. (E) The last nymphal instar prior to the adult has large obvious wing pads. (F) The adult has long wings that cover the body and enable it to fly.

Gradual Metamorphosis

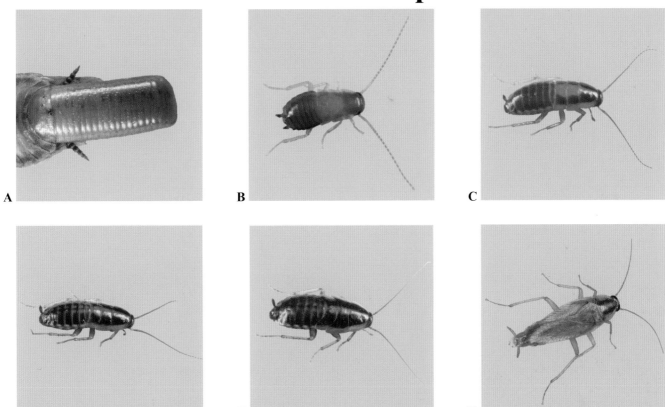

Fig. 64   Life cycle of the German cockroach (Family Blatellidae). (A) The ootheca or egg case that is still attached to the abdomen of the adult female. (B) A very young nymph. (C)-(D)-(E) Little change is seen in the developing insect except for its growth in size and the development of hardly visible wing pads along the edges of the pterothorax. (F) An adult male German cockroach.

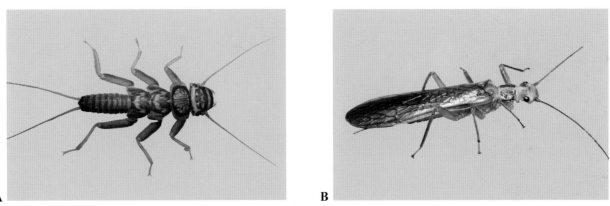

Fig. 65   The immature stages of stoneflies (Order Plecoptera) are passed in water. The adults are not aquatic but are typically found associated with streams and brooks. (A) The mature aquatic nymph of a stonefly. (B) A stonefly adult.   *Photos courtesy BP Stark*

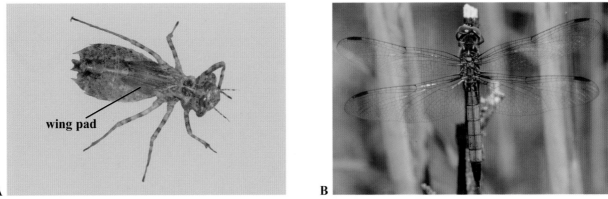

Fig. 66   The nymphs of dragonflies and damselflies (Order Odonata) are also aquatic. The adults are seen near ponds and other bodies of water. (A) The mature aquatic nymph of a dragonfly. (B) An adult dragonfly.   *Photos courtesy SW Dunkle*

Gradual Metamorphosis

Fig. 67   The life cycle of the short-winged mole cricket is seen here from egg to adult. Once the first instar nymph has hatched from the egg, the appearance is basically the same except for size. As a brachypterous species of mole cricket, the adult on the far right shows very small abbreviated wings. Family Gryllotalpidae

Fig. 68   This cluster of milkweed bugs shows almost all the stages of the life cycle. Larger nymphs have black wing pads that will eventually turn into the full-sized wings of the adults. Family Lygaeidae

Holometabolous Development

The most involved and complex insect growth pattern is called **holometabolous development** or **complete metamorphosis**. Most of you are familiar with the life cycle of a butterfly (Fig. 69) that starts with an egg, progresses through the caterpillar stages, forms a pupa (chrysalis), then eventually emerges as a butterfly. These same distinct four stages of **egg**, **larva**, **pupa**, and **adult** are seen in other advanced insect groups as well, including beetles (Fig. 70), bees and wasps (Fig. 71), fleas, flies (Fig. 72), and the nerve-winged insects.

The larval stages of insects that undergo complete metamorphosis exhibit a variety of shapes. Some have legs and are highly mobile, while others are legless and incapable of all except the slightest locomotion. Larvae have been classified into the following types based on their shape and anatomy.

Complete Metamorphosis

A   B   C   D

Fig. 69   The monarch butterfly is representative of the holometabolous development exhibited by the Order Lepidoptera. Female monarchs lay single finely sculptured eggs (A) on the host plant which is milkweed. The egg hatches into a small brightly colored caterpillar (B) or larva. Several stages are passed as caterpillars that do little else other than eat. The mature caterpillar spins a small silken pad on a leaf or twig and then hangs upside down prior to molting to a pupa. A smooth, jade green chrysalis (C) represents the pupal stage of the monarch. The outline of the adult wing is visible along its side. After one to two weeks the adult butterfly (D) emerges from the chrysalis. Family Danaidae

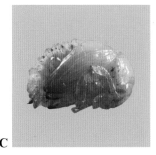

A   B   C   D

Fig. 70   The Colorado potato beetle lays its eggs (A) in clusters on plants like potatoes and eggplants. Small pinkish to red larvae hatch and immediately begin to feed on the plant foliage. The larva (B) passes through several instars before it is ready to pupate. The pupa (C) is not suspended but made among the debris of the ground. Its yellow-orange color is similar to that of the eggs. The developing wings, antennae, and legs can clearly be seen as distinct appendages on the pupa. The adult Colorado potato beetle (D) is convex and robust. Family Chrysomelidae

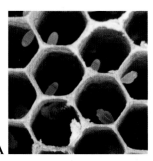

A   B   C   D

Fig. 71   The yellowjacket wasp is representative of a social insect that undergoes complete metamorphosis. The eggs (A) are laid singly inside of specially constructed cells in a honeycomb-like structure. Each egg hatches into a tiny larva (A) which is fed by the adult wasps that forage for food and bring it back to the nest. Legs on the yellowjacket larvae are lacking since they spend the entire larval period inside the same cell. When the larva is mature (B) it fills up the entire cell area. It spins a silken cap over the cell prior to pupating. The yellowjacket pupa is initially white, but develops adult colors with time. This pupa (C) has been removed from its cell. Nearly an adult, all the structures are clearly developed. The adult (D) chews its way out of the capped cell and soon becomes a functioning member of the nest. Family Vespidae

A   B   C   D

Fig. 72   Flies also undergo complete metamorphosis, most passing through the four typical stages. The female bronzebottle fly may lay hundreds of eggs (A) on carrion. The tiny larvae or maggots are very worm-like and form swarming masses in their food source. The mature larva (B) often is yellow or white and tapers towards the head. In the maggot pictured here, some of the dark food recently ingested can be seen though the thin body wall. The actual pupa forms inside the last larval skin which shrinks and hardens at the time of pupation. This hard brown oval structure is called a puparium (C). The adult fly (D) emerges from the puparium with tiny wrinkled wings which must expand and harden. Although incapable of immediate flight, the newly emerged adults run quite readily. Family Calliphoridae

Types of Larvae

Scarabaeiform - A grub-like larva that has six thoracic legs but no abdominal prolegs. Locomotion is very limited. Larvae tend to be white or light-colored. Examples are found in certain families of beetles such as the scarabs (Fig. 73) and weevils.

Vermiform - A larva that is completely legless and resembles a worm or maggot. There may be a distinct head such as in flea or wasp larvae (Fig. 75), or the head area may not be recognizable or defined as in some fly larvae (Fig. 74).

Elateriform - A long cylindrical larva that has a hard body and short legs. This type of larva is found in some beetle families such as the click beetles (Fig. 76) and darkling beetles.

Campodeiform - These are typically active and sometimes fast-moving larvae that are somewhat flattened in shape. The thoracic legs are well developed and the antennae and cerci are usually easily visible. Examples are found in many beetle families including the ground beetles, dermestids (Fig. 77), rove beetles (Fig. 78), and ladybird beetles.

Eruciform - These larvae are caterpillars or caterpillar-like. Their features include six thoracic legs and various pairs of abdominal prolegs that are used for locomotion. Examples of eruciform larvae are found in the immature stages of butterflies and moths (Fig. 79), sawflies (Fig. 80), and scorpionflies.

Types of Larvae

Fig. 73 The scarabaeiform larva is the typical white grub. It has a brown hardened head capsule and functional legs, although it is not extremely mobile and lives in the soil.

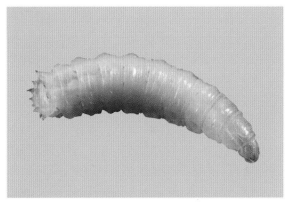

Fig. 74 Vermiform larvae are legless and worm-like. The most commonly recognized example are maggots or the larvae of flies. Many such as this one lack a clearly defined head.

Fig. 75 The legless wasp larva, although often called a grub, is also an example of a vermiform larva.

Fig. 76 This click beetle larva exemplifies the elateriform larval body. It is long, slender, cylindrical, and hardened. The legs are very short and located up near the head.

Fig. 77 This skin beetle larva shows the campodeiform body type. It is flattened and fast-moving, with cerci present.

Fig. 78 The rove beetle larva is also of the campodeiform body type. The well developed thoracic legs enable it to move quickly and act as an effective predator.

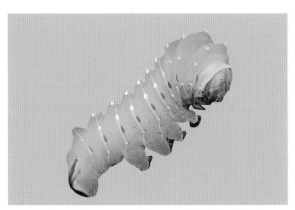

Fig. 79 Caterpillars exemplify the eruciform larval body. The six thoracic legs are sharp and pointed while locomotion is accomplished by varying numbers of abdominal prolegs.

Fig. 80 Sawfly larvae are very similar to caterpillars and are also classified as eruciform. They are not flies but belong to the same order as the wasps and bees.

The pupal stage of holometabolous insects is not actually a 'resting stage' as it is often described, but more of a reorganizational stage. During this time, adult features develop that are often radically different from the larval structures. These adult features include antennae, wings, legs, and mouthparts. They originate from clusters of cells that were present in the larva but remained dormant until the pupa. Since the wings of holometabolous insects develop internally, such insects are also described as **endopterygotes**. The pupae themselves vary considerably too, and are classified due to their characteristics. Most are effectively immobile, but some such as mosquito pupae can move both efficiently and rapidly. The pupa is a very vulnerable stage and is often attacked by parasites and predators when they can find them. For these reasons, pupae are often hidden, camouflaged, or guarded by some kind of outer protective layer provided by the larva before it molts.

Types of Pupae

Exarate - A pupa whose appendages (legs, antennae, wings) are free from the body and easily visible. In the early portion of the pupal period, exarate pupae look like white or yellow 'cadavers' of the adults. Most insect orders have exarate pupae except for butterflies and moths and some flies. (Figs. 81-83)

Obtect - A pupa whose appendages are glued tightly to the body. The pupa may have a smooth surface and be found within a protective cocoon (such as in some moths), or it may be sculptured and roughly textured (as in many butterflies). Examples of obtect pupae are also found in certain groups of flies, such as mosquitoes. (Figs. 84-86)

Coarctate - A pupa surrounded by the hardened skin of the last larval instar. The appearance is usually dark brown and cylindrical, closely resembling a rodent dropping. The hardened larval skin is termed a **puparium** and occurs in certain families of flies. (Figs. 87-89)

Hypermetamorphosis

In some parasitic insect groups that undergo complete metamorphosis, the larval type changes from the first to later instars. After hatching from the egg, the first instar larva is typically campodeiform and highly mobile. Its purpose is to find the correct host where it can complete its development. Once the host has been found and successfully parasitized, the active first instar larva molts to an immobile scarabaeiform or vermiform larva. When larval development is complete, a pupa is made as in other holometabolous insects. Examples of insects that undergo hypermetamorphosis are blister beetles (Fig. 90), mantidflies (Fig. 91), twisted-winged parasites (Fig. 92), and some flies and wasps.

Fig. 81 The exarate pupa of a carrion beetle. Family Silphidae

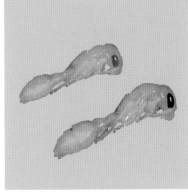

Fig. 82 The exarate pupae of *Pseudomyrmex* ants. Family Formicidae

Fig. 83 The exarate pupa of a yellowjacket wasp. Family Vespidae

Fig. 84 The obtect pupa of a corn earworm moth. Family Noctuidae

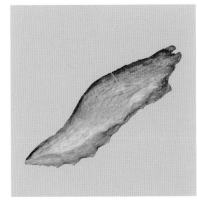

Fig. 85 The camouflaged chrysalis of the black swallowtail butterfly is also an obtect pupa. Family Papilionidae

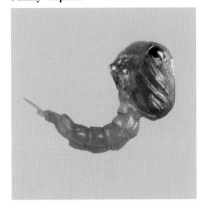

Fig. 86 The obtect pupa of a mosquito is called a tumbler, and is one of the most active of insect pupae. Family Culicidae

Fig. 87 The coarctate pupa of a flesh fly. Family Sarcophagidae

Fig. 88 The coarctate pupa of the lesser house fly is covered with rows of projections. Family Muscidae

Fig. 89 The coarctate pupa of the holarctic blow fly. Family Calliphoridae

Fig. 90 The blister beetle undergoes hypermetamorphosis with larvae that parasitize bee or grasshopper eggs. Family Meloidae

Fig. 91 The mantispid or mantidfly develops by hypermetamorphosis with larvae that parasitize spider egg cases, or the larvae of wasps and bees. Family Mantispidae

Fig. 92 The twisted-winged parasite is a small uncommon insect that undergoes hypermetamorphosis. The winged male is pictured. The wingless female and larvae parasitize wasps and hoppers. Order Strepsiptera

Taxonomic Terminology *

You have now learned to identify and recognize most of the morphological characters commonly used in the identification of insects. However, some additional vocabulary will allow you to use dichotomous keys with much greater ease and less confusion. The following terms will help you to locate or indicate structures on the insect body with greater precision and accuracy.

anterior - Referring to the front portion or head end of the body.

posterior - Referring to the tail portion or rear end of the body.

dorsal - Referring to the top side or back of the body.

ventral - Referring to the bottom side or belly of the body.

pleural - Referring to the sides of the body.

proximal - Close to or originating near the body.

distal - Away from or originating distant from the body.

*   A very thorough list of entomological terms can be found in *A Glossary Of Entomology* by J.R. de la Torre-Bueno.

Taxonomy, Classification, and Nomenclature

In order for scientists to be able to identify and discuss organisms intelligently, an organizational system for classifying and naming creatures had to be invented. The importance of such a system is clearly evident when we consider that there may be as many as 30 million different species of insects on Earth. How can we keep so many species straight? What are the criteria we should use for separating one from another? Finally, how should we name new species when we discover them?

The field of classifying organisms and dividing them into groups is called **taxonomy**. The scientist that performs such work is called a **taxonomist**. Scientists group organisms into categories based on their characteristics. Those creatures that have only a few general or basic characteristics in common are placed in very broad categories. For example, because fish and birds both have backbones they are both members of the group or category called vertebrates. As the shared traits of organisms become more defined and greater discrimination between categories is exercised, the individuals are placed in increasingly smaller categories in step-wise fashion. This is known as a **hierarchical system of classification**.

The broadest category or grouping to which an organism belongs is called a **kingdom**. For many years, biologists recognized only two kingdoms (Plants and Animals). However, three other kingdoms have been added (Monera, Protista, and Fungi), so that now there are a total of five. Some authorities may recognize even more. The next subgroup below the kingdom level is the **phylum** (plural=**phyla**). Insects belong to the Phylum Arthropoda which means 'jointed foot'.

The phylum level is only one step below the kingdom level and is still very broad. There are many other animals in the Phylum Arthropoda that are not insects. Phyla are further subdivided into **classes**. Insects make up only one of the classes in the Phylum Arthropoda. The class name is Insecta, although older reference books may show it as Hexapoda. Examples of arthropod classes that are not insects are the Arachnida (spiders), the Chilopoda (centipedes), the Diplopoda (millipedes), and the Crustacea (crustaceans). We will cover the additional classes of arthropods in the next section.

The next major subgroup is the **order**. There are 26-30 orders in the Class Insecta, depending on which taxonomic authority you follow. For the purposes of this guide, we shall follow a system with 30 different orders. Examples of insect orders are the Diptera (flies), the Mantodea (mantids), the Lepidoptera (butterflies and moths), and the Hemiptera (true bugs), just to name a few.

Taxonomic orders are composed of **families**. For example, if we select the Order Diptera or flies, we see that the mosquitoes belong to the Family Culicidae while the house fly is a member of the Family Muscidae. Some orders may have only a few families, while others like the beetles (Order Coleoptera) may have over a hundred.

The next to last taxonomic group or unit in which organisms are placed is called a **genus** (plural=**genera**). Members within a genus are extremely similar individuals that can only be further separated into **species**. A species is the most distinct category of organisms and represents individuals that can interbreed among themselves and produce viable offspring. A species name consists of two words that are underlined or written in italics. The first word is capitalized and is the name of the genus. The second word is the **species epithet** and is written entirely in lower case letters. Sometimes you will see a person's last name follow the species epithet. This is the **author** or person who originally described and classified the organism. If the author's name is in parentheses, it indicates that this species is not in the same genus in which the author originally placed it.

The two words that make up a species name are known as a **Latin binomial**. They are selected according to a set of regulations used by taxonomists and called the **Rules of Zoological Nomenclature**. It is a format that was established in 1758 by the Swedish naturalist Carolus Linnaeus. Linnaeus named so many organisms that often his name is abbreviate as 'L.' when used in the author's position. The taxonomic groups (kingdom, phylum, etc.) that we have discussed are called **taxa** (singular=**taxon**) in general. Other taxa than the ones mentioned in this guide exist. For example, a **subphylum** contains individuals that are more similar than all the members of a phylum, but that have less shared characteristics than all the members of a given class. The main taxa students need to be familiar with are the kingdom, phylum, class, order, family, genus, species. (A mnemonic phrase to remember it by is "King Phillip came over from good Spain".) Two examples of common arthropods with the names of the taxa filled in are as follows.

Common Name:	Yellow Fever Mosquito	**Common Name:**	Black Widow Spider
Kingdom:	Animal	**Kingdom:**	Animal
Phylum:	Arthropoda	**Phylum:**	Arthropoda
Class:	Insecta	**Class:**	Arachnida
Order:	Diptera	**Order:**	Araneae
Family:	Culicidae	**Family:**	Theridiidae
Genus:	*Aedes*	**Genus:**	*Latrodectus*
Species:	*aegypti*	**Species:**	*mactans*

In this manual we will use structural features and external morphology as the primary means of determining an insect's identification. However, it should be noted that other methods are available in insect identification, especially when it becomes necessary to distinguish between closely related species. For example, a recording in the form of a sonogram of a male cricket's mating song has been used in differentiating between similar appearing species of crickets. Chemical extracts of the insect cuticle have been analyzed by gas chromatography to show differences between similar species of mosquitoes, and between the commercial honey bee and the Africanized honey bee. DNA analysis is also now being used in insect species differentiation with wide applications in forensic entomology.

Many insects have common names, while others have none. Sometimes common names are widely recognized and accepted and everyone understands which insect is being discussed. In other cases however, a common name can vary from one locale to the next, or in the case of some pests, from one commodity to the next. This variation makes it extremely confusing to discuss a species unless the Latin binomial or scientific name is given. There is only one Latin name that is given to each species. Thus, if the species name *Helicoverpa zea* is used, everyone will understand to which insect it refers. However, if one farmer discusses the bollworm, and another calls it the corn earworm, and a third refers to it as the tomato fruitworm, they would all be right because those are three common names for the same insect whose Latin name is *Helicoverpa zea*.

The behavior of insects is a fascinating field of study, especially with regards to visually-oriented defense mechanisms. Pictured above is a tropical mantid in an intimidation display. Although barely an inch long, it weaves back and forth in a menacing manner, waving its spiny front legs. The hindwings, which are normally concealed, are brightly pigmented and add to the startling aspect of the display.

Phylum Arthropoda

Arthropods are made up of an incredibly wide variety of organisms of which the insects make up only a single taxonomic class. Arthropods range in size from microscopic species to those that attain a length of several meters. They are found in almost every conceivable and known type of environment and habitat.

Arthropods share certain characteristics that collectively distinguish them from all other phyla. First, they have **bilateral symmetry**. This means that only a single plane could be passed through the organism to result in mirror images on either side. Another way of thinking about it is that if a line were drawn down the center of the organism longitudinally, the left half or side would be a mirror image of the right side. Second, arthropods have paired, jointed appendages. This is easily remembered when one considers the derivation of the phylum name from the Greek (*arthro* = joint, *poda* = foot). Third, they are segmented, with the body usually defined into clearly recognizable regions or tagmata. Fourth, arthropods have a chitinous exoskeleton giving most a hard, shell-like exterior. The two remaining common features of this phylum deal with internal anatomy. One is the presence of a dorsal blood vessel. The other is a nervous system characterized by a ventral nerve cord.

Several classifications of the Phylum Arthropoda have been proposed by different taxonomists. We shall follow the system that divides the Arthropoda into three subphyla. The first is the **Subphylum Trilobita**, a group of creatures known only from fossils and called trilobites (Fig. 93). The two **extant**, or existing subphyla are separated according to the features of the mouthparts. The **Subphylum Mandibulata** has mouthparts featuring mandibles and a head with one or two pairs of antennae. This subphylum includes the insects, centipedes, millipedes, and crustaceans. The remaining group consists of chelicerate arthropods belonging to the **Subphylum Chelicerata**. The mouthparts of this taxon are fang-like or resemble pincers. They include the arachnids and the horseshoe crabs.

The dichotomous key presented at the end of this section leads to the identification of six of the most common or important classes of arthropods. It does not deal with those that the student is unlikely to encounter, or that are known only from fossil specimens. The major characters used by the key are the number of pairs of legs, the number of antennae, and the number of body regions.

Phylum Arthropoda
Insects, Spiders, Crustaceans & Others

Characteristics of Arthropoda

Jointed appendages occurring in pairs.
Body divided into distinct regions.
Bilaterally symmetrical.
Body often segmented (but not always).
Outer surface of body often hard and
shell-like (forming an exoskeleton).

Fig. 93   Trilobites are an extinct subphylum
of arthropods that are known only from
fossil specimens like this.

Subphylum Chelicerata
Arachnids and Horseshoe Crabs

Characteristics of Chelicerata

No antennae.
Two body regions (cephalothorax and abdomen).
Four pairs of walking legs.
Mouthparts consist of fangs or pincers (chelicerae)(Fig. 94).
Appendages (pedipalps) between the chelicerae and
the first pair of walking legs may be claw-like or feeler-like.

Fig. 94   Paired fangs or pincers
called chelicerae are one of the
primary features of chelicerates.

Class Arachnida   *Spiders, Scorpions & Their Relatives*

Legs usually long and obvious.
Legs usually easily visible from above or the side.
Pedipalps can be claw-like, feeler-like, or palp-like.

Class Xiphosura   *Horseshoe Crabs*

Body formed into a round-oval shell.
Long spike-like tail protrudes from back.
Edge of shell smooth in front and spiny in back.
Legs not visible from above or the side.
Adults usually 1-1.5 feet in length.
Marine animals, often found on beaches.

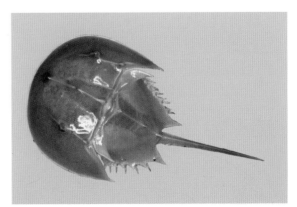

Fig. 95   The horseshoe crab is a primitive chelicerate
sometimes seen in high numbers on beaches.

41

Subphylum Mandibulata
Insects, Centipedes, Millipedes & Crustaceans

Characteristics of Mandibulata

One or two pair of antennae.
Body segmented.
Primary mouthparts are mandibles, not paired fangs or pincers.
Number of legs and body regions is variable.

Class Insecta - *Insects*   (Figs. 96-98)

Three pairs of legs.
One pair of antennae.
Segments grouped into three body regions.
Wings may or may not be present.

Class Chilopoda - *Centipedes*   (Figs. 99-101)

Fifteen or more pairs of legs.
One pair of legs per body segment.
First pair of legs thick, curved, and sharp (poison fangs).
Posterior legs at tip of trunk longer and thicker than others.
One pair of antennae present.
Body long and flattened.

Class Diplopoda - *Millipedes*   (Figs. 102-104)

Fifteen or more pairs of legs.
Two pairs of legs per most body segments.
One pair of short antennae.
Body long and often cylindrical.

Class Crustacea - *Crustaceans*   (Figs. 105-107)

Five to fourteen pairs of legs.
Two pairs of antennae (one pair often much smaller than the other).
Almost entirely marine or aquatic forms.

Order Isopoda - *Sowbugs/Pillbugs*   (Fig. 105)
Body dorso-ventrally flattened.
Seven pairs of leg-like appendages.
Only group with terrestrial crustaceans.

Order Amphipoda - *Scuds*   (Fig. 106)
Body laterally compressed.
Seven pairs of leg-like appendages.
Freshwater forms, but also found in damp situations.

Order Decapoda - *Crabs, Lobsters, Crayfish & Shrimp*   (Fig. 107)
Five pairs of leg-like appendages.
Front legs often end in a large claw.
Shell-like carapace covers the cephalothorax.

Fig. 96   The silverfish is a primitive insect that changes little from young to adult. Class Insecta

Fig. 97   The stink bug is an insect that shows incomplete metamorphosis. Class Insecta

Fig. 98   The yellowjacket wasp is an insect that undergoes complete metamorphosis. Class Insecta

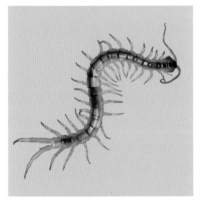

Fig. 99   Centipedes have only one pair of legs per body segment. Class Chilopoda

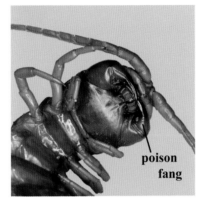

Fig. 100   The initial pair of legs on centipedes are modified into poison fangs. Class Chilopoda

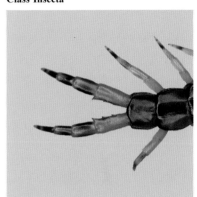

Fig. 101   The posterior legs of a centipede at the tip of the abdomen are longer and thicker. Class Chilopoda

Fig. 102   Millipedes are typically cylindrical with two pairs of legs per segment. Class Diplopoda

Fig. 103   The antennae of millipedes are shorter than centipedes and the front legs are not fangs. Class Diplopoda

Fig. 104   Some millipedes are flattened but none have a final long pair of legs. Class Diplopoda   *Courtesy FDACS/DPI*

Fig. 105   Sowbugs and pillbugs are terrestrial crustaceans found in moist conditions. Order Isopoda

Fig. 106   Scuds are extremely flattened and look like giant fleas. Order Amphipoda

Fig. 107   Aquatic crustaceans like this crab are numerous and replace insects in marine environments. Order Decapoda

Key to the Classes of Arthropoda

1A   Round to oval shell with long spike-like tail (Fig. 108). Legs only visible from below. .... **Xiphosura**

1B   Shape not as above or no spike-like tail present. Legs visible from the side and/or above. **2**

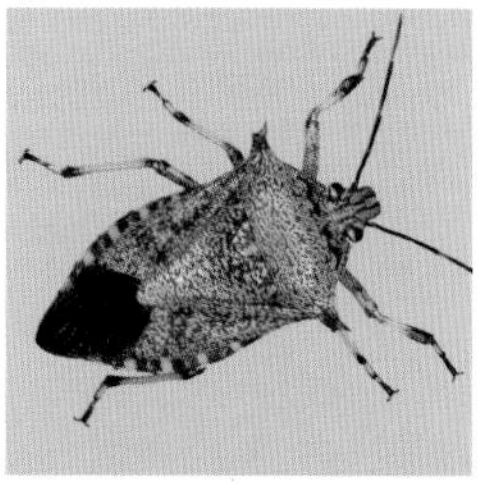

Fig. 108

2A   Three pairs of legs present. Wings may or may not be present. (Fig. 109) .... **Insecta**

2B   More than three pairs of legs present. Wings are never present. **3**

Fig. 109

3A   Four pairs of legs present. No antennae present. .... **Arachnida** (Fig. 110)

3B   Four or more pairs of legs present. One or two pairs of antennae present. **4**

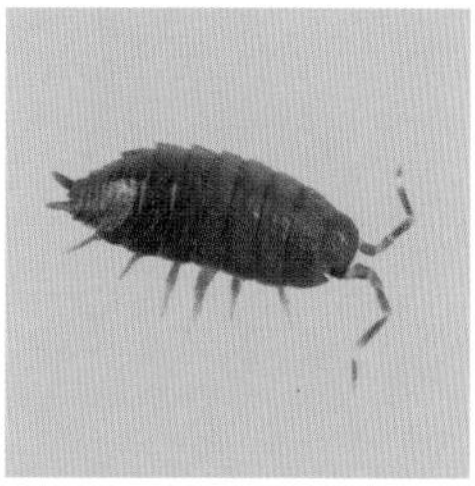

Fig. 110

4A   Less than fifteen pairs of legs present (Fig. 111). .... **Crustacea**

4B   Fifteen or more pairs of legs present. **5**

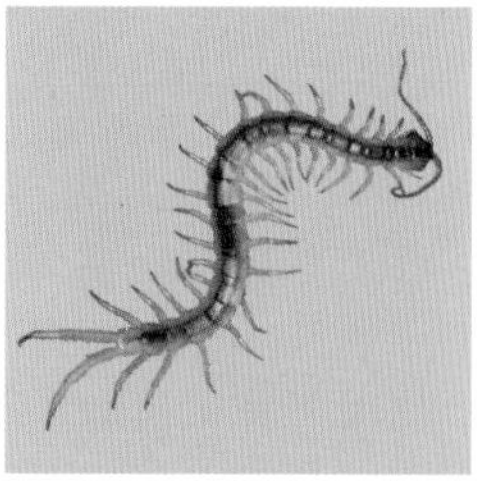

Fig. 111

5A   One pair of legs per body segment (Fig. 112). .... **Chilopoda**

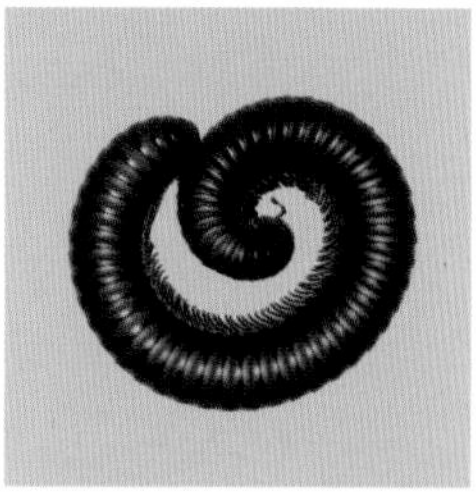

5B   Two pairs of legs per body segment (Fig. 113). ...... **Diplopoda**

Fig. 112

Fig. 113

Class Arachnida
Spiders, Scorpions, & Related Groups

Characteristics of Arachnida

Four pairs of segmented walking legs.
Two body regions or all body parts fused.
Chelicerate mouthparts (fangs or paired pincers).
No wings.
No antennae.

Terminology

chelicerae - The mouthparts of arachnids, usually consisting of fangs or pincers.
pedipalps - A pair of appendages associated with the mouth and located in front of the walking legs. Sometimes heavy and modified into claws, while other times filamentous and serve as antennae.
cephalothorax - The anterior body region formed by the combined head and thorax.
prosoma - Another term for the cephalothorax or anterior body region in some arachnids.
opisthosoma - Another term for the abdomen or posterior body region in some arachnids.
carapace - The hardened dorsal covering of the cephalothorax.
pedicel - A stalk that connects the abdomen to the cephalothorax. (Narrow in spiders.)
spinnerets - In spiders, small finger-like structures on the ventral side of the abdomen near the anus that produce silk.
telson - The region of the abdomen that is posterior to the anus.

Fig. 114　One of the two fangs that form the chelicerate mouthparts of a tarantula.
Order Araneae

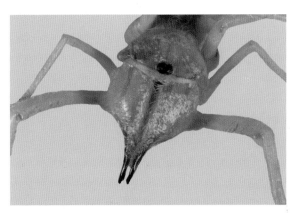

Fig. 115　Dorsal view of a camel-spider shows the top portions of the paired pincers or chelicerae.
Order Solifugae

Fig. 116　The abdomen of spiders narrows as it approaches the cephalothorax to which it is connected by a stalk called the pedicel. Order Araneae

pedicel

spinnerets

Fig. 117　At the tip of the abdomen of spiders are silk-producing structures called spinnerets.
Order Araneae

45

Class Arachnida
Spiders, Scorpions, & Related Groups

Order Araneae - *Spiders*   (Figs. 118-122)

Abdomen unsegmented.

Body usually robust and not extremely flattened.

Abdomen joined to the anterior portion of the body by a narrow stalk (pedicel).

Small finger-like structures (spinnerets) located at the tip of the abdomen.

Size and shape extremely variable.

Order Acari - *Ticks & Mites*   (Figs. 123-125)

Abdomen unsegmented.

Abdomen broadly joined or fused to the anterior region of the body.

Body usually rounded or oval.

Most ticks are flattened (unless freshly engorged).

Most mites are minute (less than a millimeter long).

Order Opiliones - *Harvestmen or Daddy-Long-Legs*   (Figs. 126-128)

All legs extremely long and thread-like.

Abdomen distinctly segmented.

Body usually oval or trapezoidal in shape.

Order Uropygi - *Whip-Scorpions or Vinegaroons*   (Figs. 129-131)

Long slender thread-like tail.

Abdomen distinctly segmented.

Pedipalps modified into heavy, thickened claws.

First pair of legs long and slender, functioning as antennae.

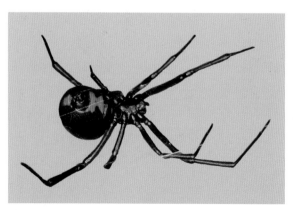

Fig. 118   A black widow spider showing the characteristic red hourglass marking on the underside of the abdomen. Order Araneae

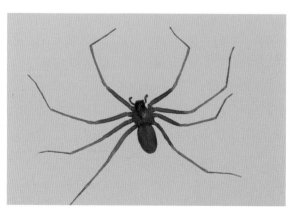

Fig. 119   The brown recluse spider has a dark fiddle-shaped marking on the carapace. Order Araneae

Fig. 120   Jumping spider.
Order Araneae

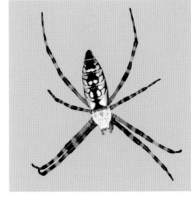

Fig. 121   Black and yellow argiope.
Order Araneae

Fig. 122   Spiny orb weaver spider.
Order Araneae

Fig. 123   Hard tick.
Order Acari

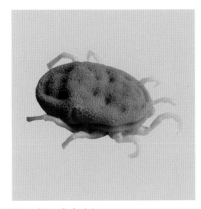

Fig. 124   Soft tick.
Order Acari

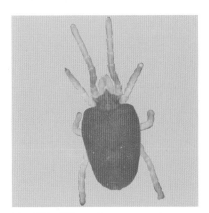

Fig. 125   Velvet mite.
Order Acari
Courtesy INHS

Fig. 126   Daddy-long-legs or harvest-
man.
Order Opiliones

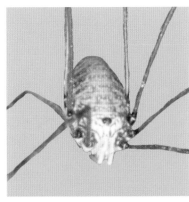

Fig. 127   Segmentation on the abdomen
of a harvestman.
Order Opiliones

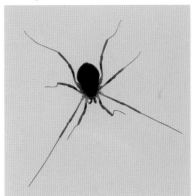

Fig. 128   Daddy-long-legs or
harvestman.
Order Opiliones

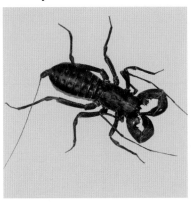

Fig. 129   Whip-scorpion or vinegaroon.
Order Uropygi

Fig. 130   Vinegaroon abdomen showing
the segmentation and slender tail.
Order Uropygi

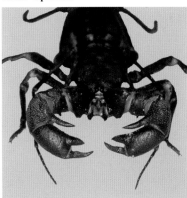

Fig. 131   Front view of a vinegaroon
showing the claw-like pedipalps.
Order Uropygi

Class Arachnida
Spiders, Scorpions, & Related Groups

Order Scorpiones - *Scorpions*   (Figs. 132-134)
Abdomen with long tail that ends in a sharp, curved stinger.
Pedipalps modified into pincer-like claws.
Abdomen distinctly segmented.

Order Pseudoscorpiones - *Pseudoscorpions*   (Figs. 135-137)
Small, usually less than 5mm in length.
Pedipalps modified into pincer-like claws.
Abdomen distinctly segmented.
Scorpion-like, but without a tail.

Order Amblypygi - *Tailless Whip-Scorpions or Whip-Spiders*   (Figs. 138-140)
Pedipalps modified into large spike-covered 'baskets' of claws.
Body round-oval and flattened.
Abdomen distinctly segmented.
First pair of legs very long and slender.

Order Solifugae - *Wind-Scorpions or Camel-Spiders*   (Figs. 141-143)
Chelicerae very large and project forward, giving head a pointed appearance.
Body elongate.
Abdomen distinctly segmented.
Pedipalps long and leg-like.
First pair of walking legs more slender than the others.

Fig. 132   Scorpion.
Order Scorpiones

Fig. 133   Stinger at the tip of a scorpion's tail.
Order Scorpiones

Fig. 134   Pincer-like claw on the pedipalp of a scorpion.
Order Scorpiones

Fig. 135   Pseudoscorpion.
Order Pseudoscorpiones

Fig. 136   Pseudoscorpion with a clutch of eggs beneath the abdomen.
Order Pseudoscorpiones

Fig. 137   Pseudoscorpion.
Order Pseudoscorpiones

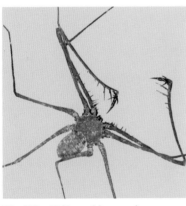

Fig. 138   Tailless whip-scorpion.
Order Amblypygi

Fig. 139   Tailless whip-scorpion.
Order Amblypygi

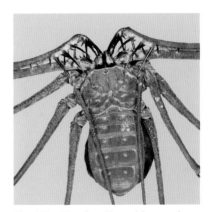

Fig. 140   Female tailless whip-scorpion with an egg mass beneath her flattened abdomen. Order Amblypygi

Fig. 141   Camel-spider or wind-scorpion.
Order Solifugae

Fig. 142   Head of a wind-scorpion showing the darkened, sclerotized chelicerae. Order Solifugae

Fig. 143   Lateral view of a wind-scorpion showing the chelicera of the left side. Order Solifugae

Key to the Orders of Arachnida

1A   Body with a long slender tail (Figs. 144-145). Front appendages with pincers or heavy claws. **2**

1B   Body without a tail. Front appendages may or may not have pincers or heavy claws. **3**

2A   Tail is obviously segmented and ends in a sharp stinger (Fig. 146). Front appendages only thickened at the ends where they form pincers. .... **Scorpiones**

2B   Tail is slender and thred-like, not obviously segmented. Front claws are massive and thickened throughout (Fig. 147). .... **Uropygi**

3A   Front appendages with pincers. Scorpion-like (Fig. 148), but usually less than 5mm. .... **Pseudoscorpiones**

3B   Front appendages without pincers. Not scorpion-like in appearance. **4**

4A   Abdomen without apparent segmentation (Fig. 149). **5**

4B   Abdomen distinctly segmented (Fig. 150). **6**

5A   Abdomen constricted at the base and joined to the anterior part of the body by a stalk. Small finger-like structures at the tip of the abdomen (Fig. 151). .... **Araneae**

5B   Abdomen broadly joined to the anterior part of the body (Figs. 152-153). No small finger-like structures at the tip of the abdomen. .... **Acari**

6A   All legs extremely long and thread-like (Fig. 154). Body usually round or oval and not flattened. .... **Opiliones**

6B   All legs not extremely long and thread-like. Body variable. **7**

7A   First pair of legs much longer than others. Front appendages are spike-covered claws (Fig. 155). ...... **Amblypygi**

7B   First pair of legs only slightly longer or more slender than others. Body elongate. Front appendages leg-like.
                                                                                    Solifugae

Fig. 144   Fig. 145

Fig. 146   Fig. 147

Fig. 148   Fig. 149

Fig. 150   Fig. 151

Fig. 152   Fig. 153

Fig. 154   Fig. 155

Class Insecta
Insects

Order Protura
Proturans

Characteristics

Minute (2mm or less)
Wings absent.
Antennae absent.
Compound eyes absent.
Styli (short paired peg-like structures) present
on the first three abdominal segments.
Mouthparts concealed within the head.
Metamorphosis is gradual.

Fig. 156   Photomicrograph of a proturan.
Order Protura
Courtesy INHS

Order Diplura
Diplurans

Characteristics

Small (7mm or less).
Wings absent.
Body long and slender.
Cerci present and variable.
Compound eyes absent.
Mouthparts concealed within the head.
Metamorphosis is gradual.

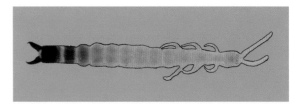

Fig. 157   Photomicrograph of a japygid dipluran. Note that
the dark outline has been added to enhance the shape.
Order Diplura

Family Campodeidae - *Campodeid Diplurans*
Cerci long (as long as the antennae) and have many segments.
Resembles silverfish in appearance, but only two caudal filaments (the cerci) present.

Family Japygidae - *Japygid Diplurans*   (Fig. 157)
Cerci short (much shorter than the antennae) and only have one segment.
Cerci resemble pincers or forceps.
Entire animal resembles a small, wingless earwig.

Order Collembola
Springtails

Characteristics

Minute (6mm or less)
Wings absent.
Body oval or elongate.
Large forked structure (furcula) at the tip of the abdomen.
Stout peg-like structure (collophore) on venter of first abdominal segment.
Antennae variable and have 4-6 segments.
Eyes present or absent.
Cerci absent.
Mouthparts concealed within the head.
Metamorphosis is gradual.

Suborder Arthropleona - *Elongate Springtails*   (Figs. 158 & 160)
Body elongate.
Abdomen with six distinct and visible segments.

Suborder Symphypleona - *Globular Springtails*   (Figs. 159 & 161)
Body is oval or globular.
Basal segments of the abdomen fused and indistinct.

Fig. 158   Photomicrograph of an elongate springtail.
Suborder Arthropleona

Fig. 159   Photomicrograph of a globular
springtail. Suborder Symphypleona
Courtesy INHS

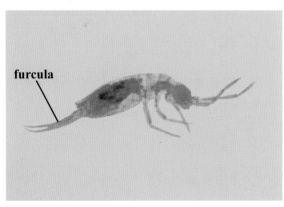

Fig. 160   Photomicrograph of an elongate springtail.
Suborder Arthropleona

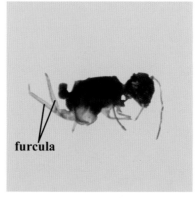

Fig. 161   Photomicrograph of a globular
springtail. Suborder Symphypleona

Order Thysanura
Silverfish and Bristletails

Characteristics

Small to medium (8-15mm).
Wings absent.
Three long filaments (two cerci and one median caudal filament) extend from the tip of the abdomen.
Body usually with scales on it.
Some abdominal segments may have short, stubby projections (styli).

Family Lepismatidae - *Silverfish & Firebrats*   (Fig. 162)

Compound eyes small and do not touch.
Tarsi with three or four segments.
Styli not present on the middle and hind coxae.

Fig. 162   Silverfish showing the flattened body form and three distinctive terminal filaments.
Family Lepismatidae

Family Machilidae - *Jumping Bristletails*   (Fig. 163)

Compound eyes large and touch.
Tarsi with three segments.
Styli present on the middle and hind coxae.

Fig. 163   Jumping bristletails have a cylindrical body form with three terminal filaments.
Family Machilidae

Order Ephemeroptera
Mayflies

Characteristics

Small to medium (5-35mm).

Four membranous wings usually present, with many veins and cross veins.

Forewings large (2-3X the size of the hind wings) and triangular in shape.

Hindwings small and rounded in shape, sometimes absent.

Tip of the abdomen with 2-3 long hair-like filaments (terminal caudal filaments).

Body elongate, slender, and soft.

Antennae short (less than the length of the head and thorax combined) and hair-like.

Mouthparts absent.

Metamorphosis is gradual.

Family Baetidae - *Small Mayflies*

Hindwings absent or extremely reduced.

Hind tarsi with three or four segments.

Compound eyes of males have noticeably larger upper facets.

Family Heptageniidae - *Stream Mayflies*

Hind tarsi with five segments.

Abdomen with two terminal caudal filaments.

Family Oligoneuridae

Venation of front wings reduced in most with
less than 10 longitudinal veins and no cross veins.

Family Caenidae

Hindwings greatly reduced or absent.

Compound eyes small and widely separated in both sexes.

Fig. 164   Dorsal view of a mayfly.
Order Ephemeroptera

Fig. 165   Lateral view of a mayfly.
Order Ephemeroptera

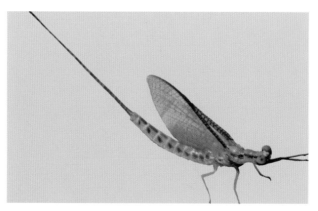

Fig 166   Oblique view of a mayfly.
Order Ephemeroptera
Courtesy INHS

Order Odonata
Dragonflies and Damselflies

Characteristics

Two pairs of long membranous wings nearly equal in size and with many cross veins.
Large spherical compound eyes that make up most of the head.
Antennae small and bristle-like (setaceous).
Chewing mouthparts.
Abdomen long and slender.
Metamorphosis is incomplete.
Immature forms are aquatic.

Terminology

stigma - Dark linear blotch along the front (costal) edge of the wing near the tip.
brace vein - Diagonal cross vein occurring below and contiguous with the proximal end of the stigma.
nodus - Small heavy cross vein near the middle of the front edge of the wing.
antenodal cross veins - Cross veins that border the front margin of the wing between the wing base and the nodus. (Referred to as **complete** if they match up directly with the cross veins below them.)
arculus - Small cross vein near the base of the wing that connects the radius and cubitus veins.
triangles - Triangular-shaped cells present near the base of the wings.
anal loop - Cluster of cells near the base of the hindwing between the anal veins.

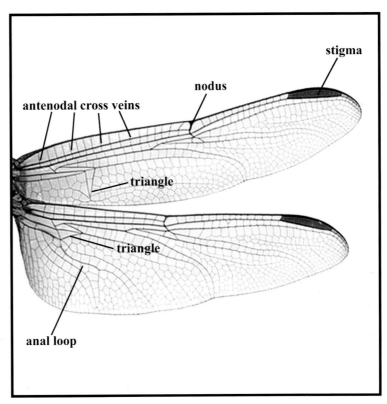

Fig. 167   Wings of a libellulid dragonfly illustrating the main taxonomic characters.
Family Libellulidae
Photo F Mitchell   Courtesy TAES

Suborder Anisoptera
Dragonflies

Characteristics

Hindwings wider at the base than the forewings.
Wings do not narrow to a stalk at the base.
Hindwings and forewings with different venation.
When alive, wings held at right angles to the body rather than together over the abdomen.
Immature forms (larvae) without visible external gills.

Family Libellulidae - *Skimmers*   (Figs. 168-170)

Anal loop is boot-shaped or foot-shaped.
Triangles in the forewings and hindwings differ in shape and location.
Eyes meet for a short distance dorsally.
Antenodal cross veins are mostly complete.
No brace vein present.

Family Aeshnidae - *Darners*   (Figs. 171-173)

Triangles in the forewings and hindwings similar in shape and location.
Eyes meet for a substantial distance dorsally.
Most antenodal cross veins incomplete.
Brace vein present.
Large dragonflies (5-8cm).

Family Gomphidae - *Clubtails*   (Figs. 174-176)

Eyes do not meet and widely separated dorsally.
Triangles in the forewings and hindwings similar in shape and location.
Most antenodal cross veins incomplete.
Brace vein present.
Some species with the last few segments of the abdomen swollen (hence the name clubtail).

Family Cordulegastridae - *Spiketails*   (Figs. 177-179)

Triangles in forewings and hindwings similar in shape and location.
Antenodal cross veins incomplete.
No brace vein present.
Eyes meet at one point dorsally.
Females with a spike-like ovipositor.

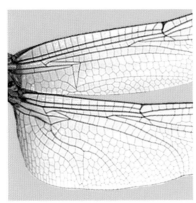

Fig. 168　Triangles in the wings of a skimmer. Family Libellulidae
Photo F Mitchell　Courtesy TAES

Fig. 169　Head and eyes of a skimmer. Family Libellulidae
Photo F Mitchell　Courtesy TAES

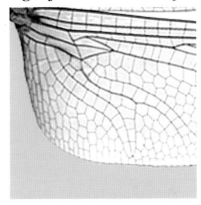

Fig. 170　Anal loop in the hindwing of a skimmer. Family Libellulidae
Photo F Mitchell　Courtesy TAES

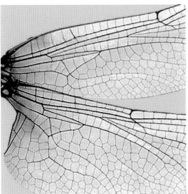

Fig. 171　Triangles in the wings of a darner. Family Aeshnidae
Photo F Mitchell　Courtesy TAES

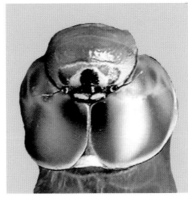

Fig. 172　Head and eyes of a darner. Family Aeshnidae
Photo F Mitchell　Courtesy TAES

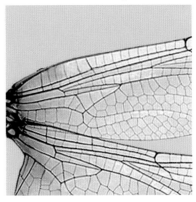

Fig. 173　Basal half of the forewing of a darner. Family Aeshnidae
Photo F Mitchell　Courtesy TAES

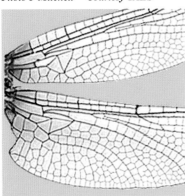

Fig. 174　Triangles in the wings of a clubtail. Family Gomphidae
Photo F Mitchell　Courtesy TAES

Fig. 175　Head and eyes of a clubtail. Family Gomphidae
Photo F Mitchell　Courtesy TAES

brace vein

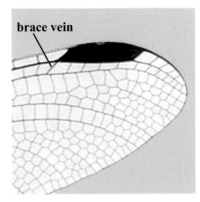

Fig. 176　Tip of the forewing of a clubtail. Family Gomphidae
Photo F Mitchell　Courtesy TAES

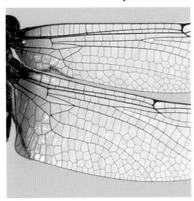

Fig. 177　Triangles in the wings of a spiketail. Family Cordulegastridae
Photo F Mitchell　Courtesy TAES

Fig. 178　Head and eyes of a spiketail. Family Cordulegastridae
Photo F Mitchell　Courtesy TAES

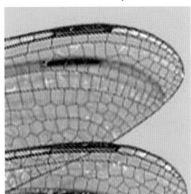

Fig. 179　Tip of the forewing of a spiketail. Family Cordulegastridae
Photo F Mitchell　Courtesy TAES

Suborder Zygoptera
Damselflies

Characteristics

Hindwings and forewings similar in shape, size, and venation.
Wings narrow towards the base, forming a 'stalk' in some families.
When alive, wings held together over the slender abdomen, or only separated slightly.
Immature forms with three visible oval flattened gills protruding from the tip of the abdomen.

Family Coenagrionidae - *Pond Damsels*   (Figs. 180-182)

Wings narrow towards the base to form a distinctive stalk.
Two antenodal cross veins present.
M3 vein begins closer to the nodus than to the arculus.
When alive, wings held folded together over the abdomen.

Family Lestidae - *Spreadwings*   (Figs. 183-185)

Wings narrow towards base to form a distinctive stalk.
Two antenodal cross veins present.
M3 vein begins closer to the arculus than to the nodus.
When alive, wings held slightly separated over the abdomen.

Family Calopterygidae - *Broad-Winged Damsels* (Figs. 186-188)

Wings do not form a stalk at the base.
Ten or more antenodal cross veins present.
When alive, wings held folded together over the abdomen.
Some species with dark wings and bright metallic bodies.

Fig. 180   Dorsal view of a pond damselfly.
Family Coenagrionidae
Photo F Mitchell   *Courtesy TAES*

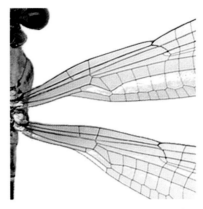

Fig. 181   Basal portion of the wings of a pond damselfly.
Family Coenagrionidae
Photo F Mitchell   *Courtesy TAES*

Fig. 182   Lateral view of a pond damselfly.
Family Coenagrionidae
Photo F Mitchell   *Courtesy TAES*

Fig. 183   Dorsal view of a spread-winged damselfly.
Family Lestidae
Photo F Mitchell   *Courtesy TAES*

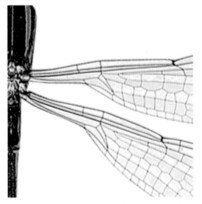

Fig. 184   Basal portion of the wings of a spread-winged damselfly.
Family Lestidae
Photo F Mitchell   *Courtesy TAES*

Fig. 185   Lateral view of a spread-winged damselfly.
Family Lestidae
Photo F Mitchell   *Courtesy TAES*

Fig. 186   Dorsal view of a broad-winged damselfly.
Family Calopterygidae
Photo F Mitchell   *Courtesy TAES*

Fig. 187   Basal portion of the wings of a broad-winged damselfly.
Family Calopterygidae
Photo F Mitchell   *Courtesy TAES*

Fig. 188   Lateral view of a broad-winged damselfly.
Family Calopterygidae
Photo F Mitchell   *Courtesy TAES*

Key to the Families of Odonata

1A   Base of forewing and base of hindwing not similar in shape. Base of forewing is much narrower than the base of hindwing. (**Suborder Anisoptera**) 2

1B   Base of forewing and base of hindwing very similar in shape, either with a stalk or gradually narrowing to the point of attachment. (**Suborder Zygoptera**) 5

2A   Compound eyes do not meet or touch when viewed dorsally. Some species with the tip of the abdomen swollen. .... **Gomphidae**

2B   Compound eyes meet or touch either at a point or for a considerable distance when viewed dorsally. Tip of abdomen not swollen. 3

3A   Triangles in forewing and hindwing different in shape and location. Anal loop is boot-shaped. .... **Libellulidae**

3B   Triangles in forewing and hindwing are similar in shape and location. Anal loop is not boot-shaped. 4

4A   Compound eyes meet only at a single point when viewed dorsally. Brace vein is absent from stigma. .... **Cordulegastridae**

4B   Compound eyes meet for a considerable distance or more than just a point when viewed dorsally. Brace vein is present at stigma. .... **Aeshnidae**

5A   Ten or more antenodal cross veins present. Wings do not form a stalk at the base, but narrow evenly to the point of attachment. .... **Calopterygidae**

5B   Two antenodal cross veins present. Wings form a stalk at the base. 6

6A   M3 vein begins closer to the nodus than to the arculus. Small triangular cell (partially formed by the M3 vein) is present below the nodus. .... **Coenagrionidae**

6B   M3 vein begins closer to the arculus than to the nodus. No small triangular cell is present. .... **Lestidae**

A giant or helicopter damselfly (Family Pseudostigmatidae) from the Amazon Basin. Found in both Central and South America, the adults of some species may have a wingspan of eight inches. They are slow deliberate flyers, navigating among the leaves and branches of the rainforests. Adults are specialized predators on spiders, taking them both from foliage and directly from their webs. *Photo courtesy SW Dunkle*

Order Mantodea
Mantids

Family Mantidae

Medium to large (20-200mm).
Triangular head with large bulging eyes.
Front legs raptorial, femur and tibia with large spines.
Pronotum extremely lengthened.
Front coxae extremely lengthened.
Chewing mouthparts.
Gradual metamorphosis.

Fig. 189   Mantid body shapes vary greatly, with some being twig-like. Family Mantidae

Fig. 190   Some mantids are extremely flattened and bark-like in their coloration. Family Mantidae

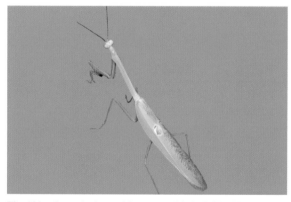

Fig. 191   In typical mantid posture, this individual holds its raptorial front legs in the classic 'praying' position. Family Mantidae

Fig. 192   The front femora and tibiae of mantids are covered with large spines. Family Mantidae

Fig. 193   Most mantids have large bulging eyes and a triangular-shaped head.   Family Mantidae

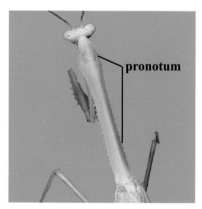

Fig. 194   An extremely long pronotum is a characteristic feature of mantids. Family Mantidae

Order Orthoptera
Grasshoppers, Crickets, and Katydids

Characteristics

Wings absent, short, or extend the full length of the body.
Leathery forewings conceal membranous hindwings.
Antennae thread-like (filiform).
Hind femora slightly to greatly enlarged.
Tarsi with fewer than five segments.
Chewing mouthparts.
Gradual metamorphosis.

Terminology

tegmen - Visible leathery forewing which is thickened and contains many veins.
tympanum - Auditory or sound-receiving structure usually characterized by an oval depression or flattened area of the body.
ovipositor - Structure found at the tip of the abdomen of females that is used for inserting eggs into a particular substrate.
stridulatory organ - Sound-producing mechanism that usually involves one structure rubbing against another.
apterous - Condition wherein the wings are totally absent.
brachypterous - Condition wherein the wings are shortened or abbreviated (usually to half the length of the abdomen or less).

Fig. 195   Short-horned grasshopper mounted with the tegmen perpendicular to the body and the colorful hindwing spread. Family Acrididae

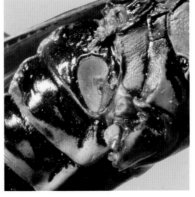

Fig. 196   The tympanum (white oval area) of a short-horned grasshopper is located on the first abdominal segment. Family Acrididae

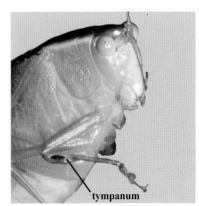

Fig. 197   The tympanum of a katydid is located on the front tibia. Family Tettigoniidae

Suborder Caelifera
Grasshoppers and Pygmy Mole Crickets

Characteristics

Antennae short (usually less than half the body length).
Tympana located on the abdomen (when present).
Tarsi with three segments or less.
Hind femora usually enlarged for jumping.
Females without an obvious visible ovipositor.

Family Acrididae - *Short-Horned Grasshoppers*   (Figs. 201-203)

Size variable (20-200mm).
Pronotum does not extend beyond the base of the wings.
All tarsi with three segments.
Tympana (when present) located on the first abdominal segment.
Hind femora usually greatly enlarged.

Family Tetrigidae - *Pygmy Grasshoppers*   (Figs. 204-206)

Less than 20mm in length.
Pronotum extends back to the tip of the abdomen.
Front and middle tarsi with two segments, hind tarsi with three segments.

Family Tridactylidae - *Pygmy Mole Crickets*   (Figs. 207-209)

Less than 10mm in length.
Hind femora enlarged and flattened.
Hind femora sometimes with semicircular imprint near joint with tibiae.
Front and middle tarsi with two segments, hind tarsi with three segments.

Fig. 198   A clutch of grasshopper eggs in soil.
Family Acrididae

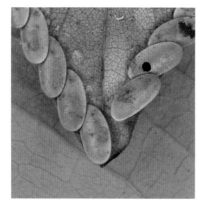

Fig. 199   A clutch of katydid eggs deposited along the edge of a leaf. Note the parasite emergence hole.
Family Tettigoniidae

Fig. 200   An exposed clutch of mole cricket eggs that are usually laid in an underground chamber.
Family Gryllotalpidae

Fig. 201   The lubber grasshopper, a species with short wings that cannot fly. Family Acrididae

Fig. 202   The American grasshopper, a short-horned grasshopper. Family Acrididae

Fig. 203   Mating short-horned grasshoppers. Family Acrididae

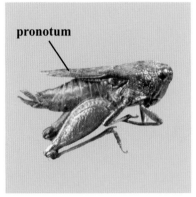

Fig. 204   Pygmy grasshopper. Note the backward-projecting pronotum that equals or exceeds the length of the body. Family Tetrigidae

Fig. 205   Pygmy grasshopper. Note the backward-projecting pronotum that equals or exceeds the length of the body. Family Tetrigidae

Fig. 206   Pygmy grasshopper. Family Tetrigidae

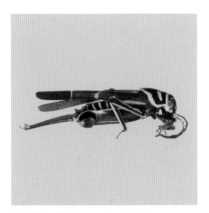

Fig. 207   Pygmy mole cricket. Family Tridactylidae

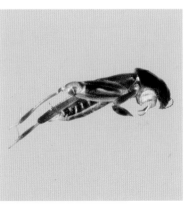

Fig. 208   Pygmy mole cricket. Family Tridactylidae

Fig. 209   Hind femur of a pygmy mole cricket. Note the semicircular shape at the apex. Family Tridactylidae

Suborder Ensifera
Crickets and Katydids

Characteristics

Antennae long (usually as long as the abdomen or longer).
Tympana located on the front tibiae when present.
Tarsi with three or four segments.
Hind femora usually moderately enlarged.
Females usually with an obvious visible ovipositor.

Family Gryllidae - *Crickets*　　(Figs. 210-212)

Small-medium (7-30mm).
Wings held flat over the body.
Tarsi with three segments.
Tympana located on front tibiae.
Ovipositor needle-like.

Family Tettigoniidae　　(*Katydids or Long-Horned Grasshoppers*)

Size variable (10-100mm).
Wings held vertically or roof-like over the body.
Tarsi with four segments.
Tympana located on front tibiae.　　(Figs. 213-215)
Ovipositor variable in shape, but often flattened and blade-like.

Family Gryllacrididae - *Camel Crickets*　　(Figs. 216-218)

Size variable (15-50mm).
Wings absent.
Tarsi with four segments.
Tympana absent.
Ovipositor variable.
Pronotum does not extend back to the abdomen (as in the shield-backed katydids).
Note: Some authorities place members of this group in the families Raphidophoridae (cave and camel crickets) and Stenopelmatidae (Jerusalem crcikets).

Family Gryllotalpidae - *Mole Crickets*　　(Figs. 219-221)

Size greater than 20mm in length.
Front legs terminate in large, finger-like claws.
Tarsi with three segments.
Antennae relatively short (less than half the body length).
Ovipositor absent.

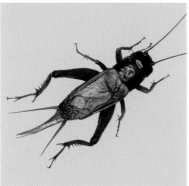

Fig. 210   Male field cricket. Note the long cerci but absence of ovipositor. Family Gryllidae

Fig. 211   Female field cricket. Note the needle-like ovipositor between the cerci. Family Gryllidae

Fig. 212   Tree cricket. Family Gryllidae

Fig. 213   A conehead katydid. Family Tettigoniidae

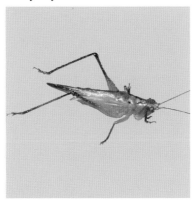

Fig. 214   Katydid or long-horned grasshopper. Family Tettigoniidae

Fig. 215   Shield-backed katydid. Note how the pronotum extends back to the abdomen. Family Tettigoniidae

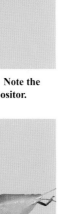

Fig. 216   Immature camel cricket. Family Gryllacrididae

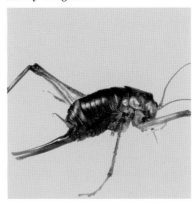

Fig. 217   Adult female camel cricket, a wingless species. Family Gryllacrididae

Fig. 218   Camel cricket. Note how the pronotum does not extend back to the abdomen. Family Gryllacrididae

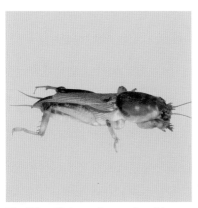

Fig. 219   Male southern mole cricket. Family Gryllotalpidae

Fig. 220   Adult short-winged mole cricket. Family Gryllotalpidae

Fig. 221   Fossorial front legs with digging claws of a northern mole cricket. Family Gryllotalpidae

Key to the Families of Orthoptera

1A   Antennae short (less than or equal to half the length of the body). **2**

1B   Antennae long (greater than half the length of the body). **5**

2A   Front legs with large hard mole-like digging claws. Medium-sized (25-40mm).
                                                                                                Gryllotalpidae

2B   Front legs without large hardened mole-like digging claws. Size variable. **3**

3A   Pronotum elongated, extending to the tip of the abdomen or beyond with the wings underneath. Small (15mm or less). Front tarsi with two segments. .... **Tetrigidae**

3B   Pronotum not elongated and does not extend to the tip of the abdomen. Size variable. Front tarsi with two or three segments. **4**

4A   Small (less than 10mm). Hind femur flattened and enlarged for jumping, sometimes with a semicircular marking near the joint with the tibia. Portions of front leg enlarged for digging. Front tarsi with two segments. .... **Tridactylidae**

4B   Size variable, but usually greater than 20mm. Hind femur enlarged and robust for jumping. Front legs not enlarged for digging. Front tarsi with three segments. .... **Acrididae**

5A   Wings held flat over the body. Tarsi with three segments. .... **Gryllidae**

5B   Wings held roof-like over the body. Tarsi with four segments. **6**

6A   Wings present, although sometimes reduced. Hind femur only moderately enlarged.
                                                                                                Tettigoniidae

6B   Wings absent. Hind femur greatly enlarged. **7**

7A   Pronotum extends back to first abdominal segment. Tympana located on front tibiae near the joint with the femur. .... **Tettigoniidae**

7B   Pronotum does not extend back to first abdominal segment. Tympana are absent.
                                                                                                Gryllacrididae

The order Orthoptera includes many fascinating insect groups. Among the most spectacular are the neotropical leaf-mimicking katydids. Pictured above is a female leaf katydid in the genus *Cycloptera* that was collected in the lowland rainforest of the Amazon Basin of Peru. There are approximately one hundred described species of leaf-mimicking katydids in the New World.

Order Blattaria
Cockroaches

Characteristics of Blattaria

Small-large (5-60mm).
Body oval and dorsoventrally flattened, with long slender legs.
Head concealed by pronotum when viewed from above.
Wings usually present (although sometimes short) and held flat over the back.
Front wings leathery (tegmina).
Antennae long (usually as long as the body) and thread-like.
Legs usually very spiny.
Cerci large and with many segments.
Chewing mouthparts.
Gradual metamorphosis.

Terminology

styli - Pair of short, finger-like structures at the tip of the abdomen below the cerci.
subgenital plate - Sternite at the tip of the abdomen.

Family Blattidae - *Blattid Cockroaches*   (Figs. 222-224)

Medium-large (20-40mm).
Spines on ventrolateral margin of front femora nearly equal in length or decrease in size distally.
Styli of males slender and similar.
Subgenital plate of female divided longitudinally.

Family Blatellidae - *Blatellid Cockroaches*   (Figs. 225-227)

Small-medium (usually less than 20mm).
Spines on ventrolateral margin of front femora margin large proximally and smaller distally.
Styli of males asymmetrical.
Subgenital plate of female not divided longitudinally.

Family Blaberidae - *Blaberid Cockroaches*   (Figs. 228-230)

Size variable. (Most 15-60mm. If smaller, then pale green in color.)
This group includes the giant cockroaches.

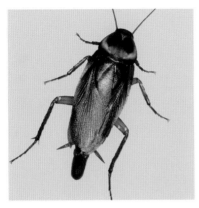

Fig. 222   Female American cockroach with eggcase.
Family Blattidae

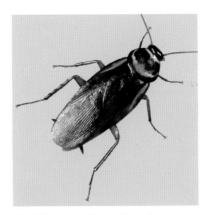

Fig. 223   Australian cockroach.
Family Blattidae

Fig. 224   Brown cockroach.
Family Blattidae

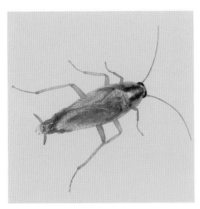

Fig. 225   German cockroach.
Family Blatellidae

Fig. 226   Brownbanded cockroach.
Family Blatellidae

Fig. 227   *Blatella lituricollis*
Family Blatellidae

Fig. 228   Giant cockroach.
Family Blaberidae

Fig. 229   Madagascar hissing cockroach.
Family Blaberidae

Fig. 230   Cuban cockroach.
Family Blaberidae

Order Phasmatodea
Walkingsticks

Characteristics of Phasmatodea

Body usually long and slender.
All legs similar, modified for walking.
All tarsi with either five segments or three segments.
Base of front leg often notched or indented.
Cerci short, consisting of only one segment.
Chewing mouthparts.
Gradual metamorphosis.

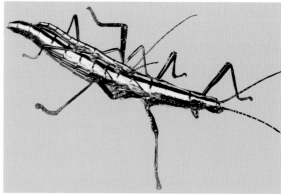

Family Pseudophasmatidae
(*Striped Walkingsticks*)

Dorsum often with a dark median longitudinal stripe
and two light lateral longitudinal stripes.
Wings absent.
Tarsi with five segments.

Fig. 231 Striped walkingsticks are sometimes found as mating pairs with the smaller male on top.
Family Pseudophasmatidae

Family Timemidae
(*Timema Walkingsticks*)

Small (15-30mm)
Wings absent.
Tarsi with three segments.
Similar to earwigs (Dermaptera) in appearance.

Fig. 232 Timemid walkingsticks are small and sometimes mistaken for earwigs.
Family Timemidae

Family Phasmatidae
(*Winged Walkingsticks*)

Short wings present.
Head with two spines on top.
First abdominal tergum longer than wide.
Tarsi with five segments.

Fig. 233 Short wings are present on the winged walkingsticks.
Family Phasmatidae

Family Heteronemeidae
(*Common Walkingsticks*)

Wings absent.
Head without spines on top.
First abdominal tergum shorter than wide.
Tarsi with five segments.

Fig. 234 Common walkingsticks are wingless and do not have spines on the top of the head.
Family Heteronemeidae

Order Isoptera
Termites

Fig. 235   A queen termite with a worker of normal size for comparison.

Characteristics of Isoptera

Small-medium (5-20mm).
Wings absent or present.
When present, there are four membranous wings, equal in size and shape.
Wings reduced in venation, especially lacking in cross veins.
Antennae bead-like.
Chewing mouthparts.
Gradual metamorphosis.

Fig. 236   The reproductive form of a termite with the wings intact.

Note: Termites are social insects with three distinct classes or castes that vary greatly in appearance even within the same species. General descriptions of these castes are provided below.

Queen

Extremely large individual (5-10X size of its nestmates in many species). Abdomen may be so distended with eggs that the dorsal sclerites appear only as small black spots on the body.

Fig. 237   The reproductive form of a termite after the wings have broken off.

Reproductive Caste

Wings present, or small scale-like wing remnants present.
Body dark-colored.

Soldier Caste

Wings or wing remnants absent.
Body light-colored.
Head usually large, darker than the body, and with large mandibles or nasus.

Fig. 238   The soldier caste of a termite.

Worker Caste

Wings or wing remnants absent.
Wing buds may be present on nymphs or immatures.
Body light-colored.
Head not overly large and is without large mandibles or nasus.

Fig. 239   The worker caste of a termite.

Terminology

caste - Morphologically distinct group of individuals within a species that performs specific functions within the nest or community of social insects to which it belongs.
fontanelle - Shallow, light-colored spot, depression, or opening on the top of the head between the eyes.
nasus - Elongated snout protruding from the head of some soldier termites.
scale - Wing remnant or stub that remains attached to the thorax of a reproductive individual after the wings have been shed.

Order Isoptera
Termites

Note: The characters given for the families below are for the reproductive caste only.

Family Kalotermitidae   *(Drywood, Dampwood, & Powderpost Termites)*

Anterior margin of wing with three or more heavy longitudinal veins.
Distal part of third longitudinal anterior vein with one or more heavy anterior branches.
Ocelli present.
Antennae with less than 21 segments.

**Fig. 240   Winged reproductive form of a drywood termite.
Family Kalotermitidae**
Photo RH Scheffrahn   *Courtesy IFAS/UF*

Family Rhinotermitidae   *(Subterranean Termites)*

Anterior margin of wing with two heavy longitudinal veins.
Second heavy anterior longitudinal vein does not branch.
Scales of front wings longer than the pronotum.

**Fig. 241   Winged reproductive form of a subterranean termite.
Family Rhinotermitidae**

Family Termitidae   *(Desert & Nasutiform Termites)*

Anterior margin of wing with two heavy longitudinal veins.
Second heavy anterior longitudinal vein does not branch.
Scales of front wings shorter than the pronotum.

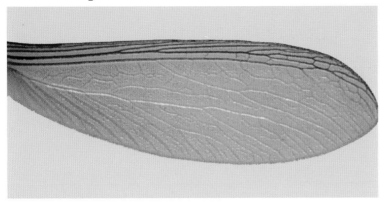

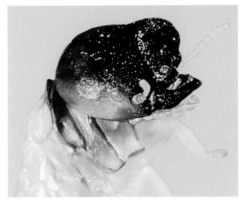

Fig. 242   Wing venation of a dampwood termite. Note there are at least three heavy longitudinal veins in the anterior margin of the wing.
Family Kalotermitidae
Photo RH Scheffrahn   *Courtesy IFAS/UF*

Fig. 243   Head of the soldier of a powderpost drywood termite.
Family Kalotermitidae
Photo RH Scheffrahn   *Courtesy IFAS/UF*

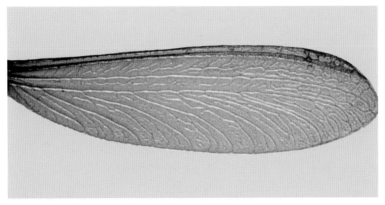

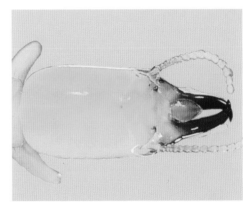

Fig. 244   Wing venation of a subterranean termite. Note the two heavy longitudinal veins in the anterior margin of the wing.
Family Rhinotermitidae
Photo RH Scheffrahn   *Courtesy IFAS/UF*

Fig. 245   Head of a soldier of a subterranean termite.
Family Rhinotermitidae
Photo RH Scheffrahn   *Courtesy IFAS/UF*

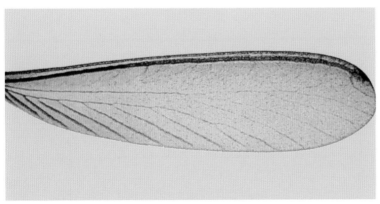

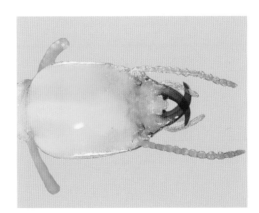

Fig. 246   Wing venation of a higher subterranean termite. Note the two heavy longitudinal veins in the anterior margin of the wing.
Family Termitidae
Photo RH Scheffrahn   *Courtesy IFAS/UF*

Fig. 247   Head of a soldier of a higher subterranean termite.
Family Termitidae
Photo RH Scheffrahn   *Courtesy IFAS/UF*

Order Plecoptera
Stoneflies

Characteristics of Plecoptera

Small-medium (5-60mm).

Body elongate, flattened, and soft.

Four membranous wings usually present and held flat over the back.

Hindwings wider than forewings, with anal area folding like a fan.

Antennae long (greater than the length of the head and thorax combined) and thread-like.

Cerci long and with many segments.

Ocelli present.

Tarsi with three segments.

Chewing mouthparts.

Gradual metamorphosis.

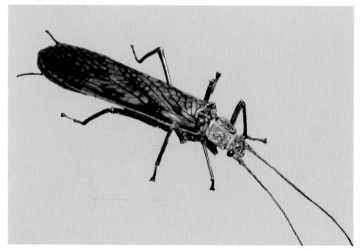

Fig. 248　Adult stonefly.
Family Perlodidae
Courtesy BP Stark

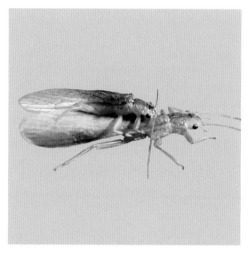

Fig. 249　Mating stoneflies.
Family Perlidae
Courtesy BP Stark

Fig. 250　Adult stonefly.
Family Pteranarcyidae
Courtesy BP Stark

Fig. 251　Dorsal view of a stonefly mounted with the wings spread.

Order Dermaptera
Earwigs

Characteristics of Dermaptera

Small-medium (5-25mm).
Body elongate and somewhat flattened.
Tip of abdomen with large pincers (cerci).
Wings present or absent.
Forewings short and leathery (referred to as either tegmina or elytra).
Forewings meet in a straight line down the back.
Hindwings membranous and rounded, folding beneath the forewings which conceal them except for the tips.
Antennae thread-like.
Tarsi with three segments.
Chewing mouthparts.
Gradual metamorphosis.

**Fig. 252   Common earwig.
Family Forficulidae**

Family Forficulidae - *Common Earwigs*

Body approximately 10-15mm long and brown.
Antennae with 12-15 segments.
Second tarsal segment lobed, extending beneath the third.

**Fig. 253   Long-horned earwig
Family Labiduridae**

Family Labiduridae - *Long-Horned Earwigs*

Antennae with 16-30 segments.
First antennal segment equal to or longer than segments 4-5-6 combined.
Second tarsal segment cylindrical and not extended.

**Fig. 254   Little earwig.
Family Labiidae**

Family Labiidae - *Little Earwigs*

Body approximately 4-8mm long.
Antennae with 11-15 segments.
First antennal segment shorter than segments 4-5-6 combined.
Second tarsal segment cylindrical and not extended.

**Fig. 255   Black earwig.
Family Chelisochidae**

Family Chelisochidae - *Black Earwigs*

Body approximately 15-20mm long and black.
Second tarsal segment narrow, extended, and has a fringe of hairs.

Family Carcinophoridae

Wings absent.
Cerci asymmetrical.
Second tarsal segment cylindrical.

**Fig. 256   Carcinophorid earwig.
Family Carcinophoridae**

Order Embioptera
Webspinners

Characteristics of Embioptera

Small (4-8mm), elongate, and somewhat flattened or cylindrical.

Wings present or absent.

All four wings membranous and approximately equal in size.

Tarsi on front legs enlarged.

Femora of hind legs enlarged.

Antennae thread-like and less than half the length of the body.

Cerci present.

Ocelli absent.

Tarsi with three segments.

Chewing mouthparts.

Gradual metamorphosis.

Fig. 257   Adult webspinner.
Order Embioptera

Order Zoraptera
Zorapterans

Characteristics of Zoraptera

Minute (1-3mm).

Antennae thread-like or bead-like, with nine segments.

Cerci present and terminate in a bristle.

Wings present or absent.

If wings present:

    Body dark-colored.

    Compound eyes and ocelli present.

If wings absent:

    Body light-colored.

    Compound eyes and ocelli are absent.

Chewing mouthparts.

Gradual metamorphosis.

Fig. 258   Photomicrograph of a zorapteran.
Order Zoraptera

Order Psocoptera
Booklice and Barklice

Characteristics of Psocoptera

Small (6mm or less).
Wings present or absent.
Forewings larger than hindwings.
Wings held roof-like over the body.
Antennae as long as the body and thread-like.
Enlarged thorax gives a humpbacked appearance.
Body soft.
Cerci absent.
Tarsi with two or three segments.
Chewing mouthparts.
Gradual metamorphosis.

Fig. 259   Adult barklouse with wings held roof-like over the body.
Order Psocoptera

Fig. 260   Cluster of winged barklice on a tree trunk.
Order Psocoptera

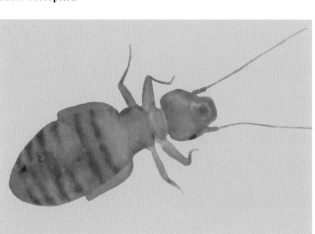

Fig. 261   Adult booklouse.
Order Psocoptera
Photo JL Castner   Courtesy IFAS/UF

Fig. 262   Immature and adult booklouse.
Order Psocoptera

Order Anoplura
Sucking Lice

Characteristics of Anoplura

Small (5mm or less).
Wings absent.
Head narrower than the thorax.
Tarsi with one segment and one claw.
Antennae short (no longer than the length of the head).
Antennae usually visible.
Compound eyes small or absent.
Cerci absent.
Sucking mouthparts.
Gradual metamorphosis.

Family Haematopinidae - *Mammal Sucking Lice* (Figs. 263-265)

Body pear-shaped or teardrop-shaped.
Eyes (or tubercles in place of eyes) absent.
Thorax markedly narrower than the widest part of the abdomen.
Hind legs similar in form and size to the front legs.

Family Pediculidae - *Human Lice* (Figs. 266-268)

Body pear-shaped or teardrop-shaped.
Eyes (or tubercles in place of eyes) present.
Thorax markedly narrower than the widest part of the abdomen.
Hind legs similar in form and size to the front legs.

Family Phthiridae - *Crab Lice* (Figs. 269-271)

Body squarish or shield-shaped (crab-like).
Eyes (or tubercles in place of eyes) present.
Thorax as wide or wider as the widest part of the abdomen.
Hind legs and mesothoracic legs larger and thicker than the front legs.

Fig. 263   Photomicrograph of a mammal sucking louse.
Family Haematopinidae

Fig. 264   Photomicrograph of a hog louse.
Family Haematopinidae

Fig. 265   Photomicrograph of a hog louse.
Family Haematopinidae

Fig. 266   Photomicrograph of a human body louse.
Family Pediculidae

Fig. 267   Photomicrograph of a human body louse.
Family Pediculidae

Fig. 268   Human body louse feeding.
Family Pediculidae

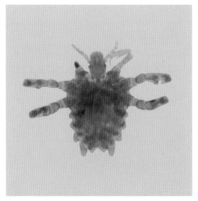

Fig. 269   Photomicrograph of a human crab or pubic louse.
Family Phthiridae

Fig. 270   Photomicrograph showing the enlarged tarsal claws of the crab louse.
Family Phthiridae
Photo JL Castner   Courtesy IFAS/UF

Fig. 271   Human crab louse attached to a pubic hair.
Family Phthiridae
Photo JL Castner   Courtesy IFAS/UF

Order Mallophaga
Chewing Lice

Characteristics of Mallophaga

Small (10mm or less).
Wings absent.
Head as wide or wider than the thorax.
Tarsi with one or two segments and one or two claws.
Antennae short (no longer than the length of the head) and often concealed in grooves.
Compound eyes present, but small.
Cerci absent.
Chewing mouthparts.
Gradual metamorphosis.

Family Menoponidae - *Bird Lice*

Antennae clubbed and usually concealed.
Head generally triangular in shape, and widens behind the eyes.
Tarsi with two claws.

Fig. 272   **Photomicrograph of a bird louse. Family Menoponidae**

Family Philopteridae - *Feather Chewing Lice*

Antennae thread-like and usually visible.
Tarsi with two claws.

Fig. 273   **Photomicrograph of a feather chewing louse. Family Philopteridae**

Order Hemiptera
True Bugs

Characteristics of Hemiptera

Two pairs of wings usually present.
Forewings modified to hemelytra, hindwings entirely membranous.
Beak arises from the anterior or front portion of the head.
Antennae obvious or hidden; 4-5 segments when obvious.
Ocelli present or absent.
Some groups (ex. stink bugs) with scent glands on the sides of the thorax.
Mouthparts consist of a piercing-sucking 'beak', with no palps.
Gradual metamorphosis.

Terminology

hemelytron - Forewing with the basal portion thickened and leathery,
and the distal portion (wingtip) membranous.
corium - Thickened basal portion of the hemelytron.
clavus - Triangular anal portion of the hemelytron that borders the scutellum.
cuneus - Apical portion of the corium that is triangular, and defined by a suture and the wingtip.
scutellum - Often triangular-shaped structure located on the dorsum between the bases of the wings.
membrane - Membranous portion of the hemelytron (the wingtip).

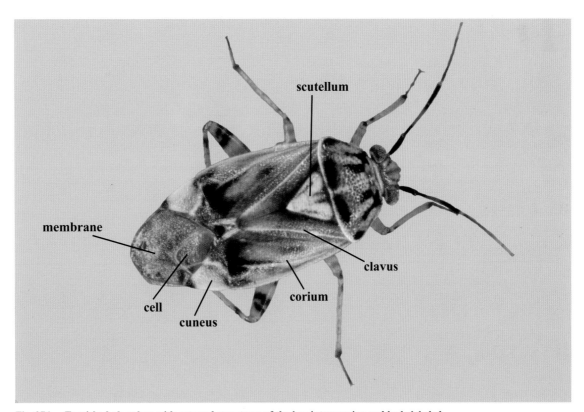

**Fig. 274   Tarnished plant bug with external structures of the hemipteran wing and body labeled.
Family Miridae**

Suborder Cryptocerata
Short-Horned Bugs

Characteristics of Cryptocerata
Antennae hidden, concealed in grooves on the head.
Ocelli absent.
Adapted for aquatic or semi-aquatic habitats.

Family Corixidae - *Water Boatmen*   (Figs. 275-277)
Hind legs oar-like with long hairs.
Dorsal surface flattened.
Dark lateral markings on dorsum and on wings.
Front tarsi scoop-shaped and surrounded by long hairs.
Space between the eyes usally greater than or equal to the width of a single eye.

Family Notonectidae - *Backswimmers*   (Figs. 278-280)
Hind legs oar-like with long hairs.
Dorsum convex and keel-like, V-shaped when viewed from the tail.
Wings unpigmented, the tips without veins.
Space between the eyes usally less than the width of a single eye.

Family Nepidae - *Waterscorpions*   (Figs. 281-283)
All legs long and slender.
Terminal appendages that form the air tube equal to or greater than half the length of the body.
Front legs raptorial, with a tooth-like spur on the femur.
Venation of membrane reticulate.
The head of some species mantid-like in appearance.

Family Belostomatidae - *Giant Water Bugs*   (Figs. 284-285)
Hind legs oar-like with long hairs. (Note: Hairs may lie flat against tibia and be difficult to see.)
Dorsum flattened and oval-broadly oval in shape.
Terminal appendages (when visible) less than half the length of the body.
Front legs raptorial.

Family Naucoridae - *Creeping Water Bugs*   (Fig. 286)
Dorsum rounded and smooth.
Front legs raptorial, the femora greatly thickened and almost triangular.
Front tibiae curved and blade-like.
Membrane without veins.
Edge of abdomen may extend beyond the wings and be banded.

Fig. 275   Dorsal view of a water boatman.
Family Corixidae

Fig. 276   Lateral view of a water boatman. Note oar-like hind legs shown projecting forwards.   Family Corixidae

Fig. 277   Head of a water boatman. Note the curvature of the eyes.
Family Corixidae

Fig. 278   Dorsal view of a backswimmer.
Family Notonectidae

Fig. 279   Lateral view of a backswimmer. Note the high keel-like dorsum.   Family Notonectidae

Fig. 280   Head of a backswimmer. Note the closely spaced somewhat bulging eyes.
Family Notonectidae

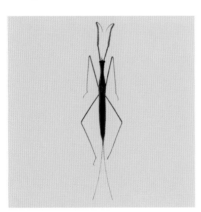

Fig. 281   Dorsal view of a slender-bodied waterscorpion.
Family Nepidae

Fig. 282   Dorsal view of an oval-bodied waterscorpion.
Family Nepidae

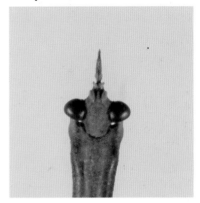

Fig. 283   Head and beak of a slender-bodied waterscorpion.
Family Nepidae

Fig. 284   Dorsal view of a giant water bug.
Family Belostomatidae

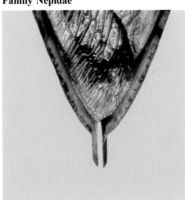

Fig. 285   The short air tubes at the tip of the abdomen of a giant water bug.
Family Belostomatidae

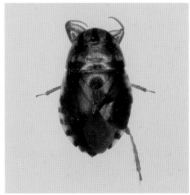

Fig. 286   Dorsal view of a creeping water bug.
Family Naucoridae

Suborder Gymnocerata
Long-Horned Bugs

Characteristics of Gymnocerata

Antennae obvious, and with four to five segments.
Ocelli present or absent.
Adapted for various habitats.

Family Pentatomidae - *Stink Bugs*   (Figs. 287-289)

Antennae with five segments.
Ocelli present.
Shield-shaped or broadly oval.
Large triangular scutellum.
Dorsum relatively flat.
Body often covered with small depressions or punctations.
Pronotum sometimes with spines at the edges.
Tibiae with either no spines, or with sparse weak spines.

Family Scutelleridae - *Shield-Backed Bugs*   (Figs. 290-292)

Antennae with five segments.
Ocelli present.
Shield-shaped or broadly oval.
Scutellum conceals the wings and covers the entire dorsum, resulting in a resemblance to beetles.
Dorsum convex and rounded.
Body usually covered with small depressions or punctations.
Pronotum never with spines at the edges.

Family Cydnidae - *Burrower Bugs*   (Figs. 293-295)

Small (7mm or less).
Antennae with five segments.
Ocelli present.
Broadly oval in shape.
Large triangular scutellum.
Tibiae densely covered with stout spines.

Family Thyreocoridae - *Negro Bugs*   (Figs. 296-298)

Small (3-6mm or less), and shiny black.
Antennae with five segments.
Ocelli present.
Round to oval in shape.
Scutellum covers most of the dorsum, concealing the wings.
Spines, if present on the tibiae, are sparse and weak.

Fig. 287   Harlequin stink bug.
Family Pentatomidae

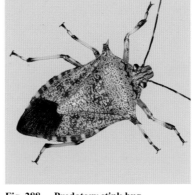

Fig. 288   Predatory stink bug.
Family Pentatomidae

Fig. 289   Predatory stink bug.
Family Pentatomidae

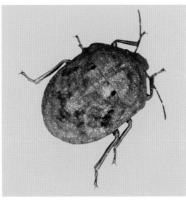

Fig. 290   Dorsal view of a shield-backed
bug.
Family Scutelleridae

Fig. 291   Lateral view of a shield-backed
bug.
Family Scutelleridae

Fig. 292   Tropical shield-backed bug.
Family Scutelleridae

Fig. 293   Dorsal view of a burrower
bug.
Family Cydnidae

Fig. 294   Lateral view of a burrower
bug.
Family Cydnidae

Fig. 295   Tibia of a burrower bug. Note
the many stout spines.
Family Cydnidae

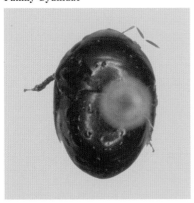

Fig. 296   Dorsal view of a negro bug.
Family Thyreocoridae

Fig. 297   Lateral view of a negro bug.
Family Thyreocoridae

Fig. 298   Tibia of a negro bug. Note the
lack of stout spines.
Family Thyreocoridae

Suborder Gymnocerata
Long-Horned Bugs

Family Reduviidae - *Assassin Bugs*   (Figs. 300-302)

Antennae with four segments.
Ocelli present (except in thread-legged assassin bugs).
Large, usually curved beak ends in a groove at the base and between the two front legs.
Transverse suture present between the eyes.
Membrane with two large closed oval cells and a single vein extending to the margin.

Family Phymatidae - *Ambush Bugs*   (Figs. 303-305)

Antennae with four segments and slightly clubbed.
Ocelli present.
Front legs raptorial with front femora greatly thickened.
Abdomen extends laterally beyond the wings.
Membrane with parallel branching veins extending to margin.

Family Lygaeidae - *Seed Bugs*   (Figs. 306-308)

Small (most less than 10mm).
Antennae with four segments.
Ocelli present.
Membrane with four or five parallel veins.

Family Coreidae - *Leaf-Footed Bugs*   (Figs. 309-311)

Medium in size (most greater than 10mm).
Antennae with four segments.
Ocelli present.
Membrane with many (15-20) parallel veins.
Some species with hind tibiae flattened and/or greatly expanded.

Family Aradidae - *Flat or Fungus Bugs*

Antennae with four segments.
Ocelli absent.
Extremely flattened, almost paper-thin in profile.
Wings (when present) cover only a small area in center of dorsum.
Membrane with reticulate to variable venation.
Brown or black in color.

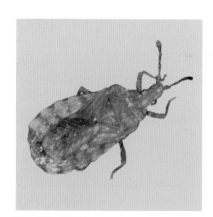

Fig. 299   Flat bug.
Family Aradidae

Fig. 300   Bloodsucking conenose bug.
Family Reduviidae

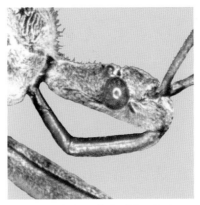

Fig. 301   Head and bek of the wheel
bug.
Family Reduviidae

Fig.. 302   Venation in the membrane of
an assassin bug's wing.
Family Reduviidae

Fig. 303   Dorsal view of an ambush bug.
Note the wide lateral extensions of
the abdomen. Family Phymatidae

Fig. 304   Head and raptorial front leg of
an ambush bug.
Family Phymatidae

Fig. 305   Head of an ambush bug
showing the slightly clubbed antennae.
Family Phymatidae

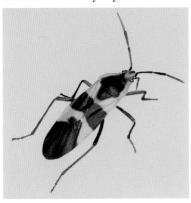

Fig. 306   Milkweed bug.
Family Lygaeidae

Fig. 307   Big-eyed bug.
Family Lygaeidae

Fig. 308   Venation in the wing of a seed
bug. Note there are only a few parallel
veins present. Family Lygaeidae

Fig. 309   Large leaf-footed bug with
thickened femora and tibiae.
Family Coreidae

Fig. 310   Squash bug. Note the hind
femora and tibiae are enlarged.
Family Coreidae

Fig. 311   Venation in the wing of a leaf-
footed bug. Note the many parallel
veins present. Family Coreidae

Suborder Gymnocerata
Long-Horned Bugs

Family Miridae - *Plant Bugs*　　(Figs. 313-315)
Antennae with four segments.
Ocelli absent.
Cuneus present on corium.
Membrane with 1-2 closed cells at base and no other veins.
Wingtips tilt down at an angle when viewed in profile.

Family Tingidae - *Lace Bugs*　　(Figs. 316-318)
Small (usually 5mm or less).
Body moderately to extensively covered with lacy, reticulate, sometimes bubbly appearing sculpturing.
Antennae with four segments and slightly clubbed.
Ocelli absent.

Family Nabidae - *Damsel Bugs*　　(Figs. 319-321)
Antennae with four segments.
Ocelli present.
Membrane with many small cells around the margin.
Usually slender and elongate in shape.

Family Pyrrhocoridae - *Red Bugs or Stainers*　　(Figs. 322-324)
Antennae with four segments.
Ocelli absent.
Membrane with 2-3 closed cells at base and 6-10 long radiating veins.
Brightly colored.

Family Cimicidae - *Bed Bugs*　　(Fig. 312)
Antennae with four segments.
Ocelli absent.
Wings absent.
Flattened (unless engorged), and round to oval in shape.
Anterior edge of pronotum wraps around head.
Abdomen bears lateral rows of hairs.
Brown or black in color.

**Fig. 312　Dorsal view of a feeding bed bug. Note the blood seen through the body.
Family Cimicidae**

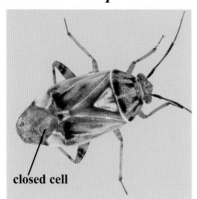

Fig. 313   Tarnished plant bug. Note the closed cell in the wing membrane. Family Miridae

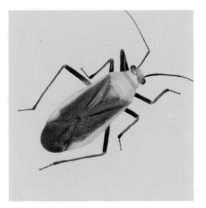

Fig. 314   Dorsal view of a common plant bug. Family Miridae

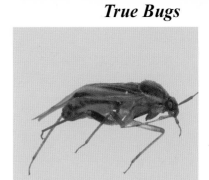

Fig. 315   Lateral view of a plant bug showing how the wingtip tilts down at an angle. Family Miridae

Fig. 316   Avocado lace bug. Family Tingidae *Courtesy FDACS/DPI*

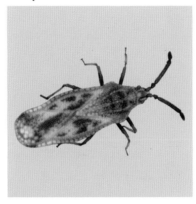

Fig. 317   Lantana lace bug. Family Tingidae

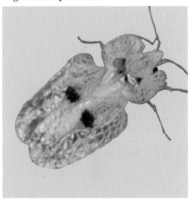

Fig. 318   Sycamore lace bug. Family Tingidae

Fig. 319   Dorsal view of a damsel bug. Family Nabidae

Fig. 320   Dorsal view of a damsel bug. Family Nabidae

Fig. 321   Venation in the wing of a damsel bug. Note the many small cells along the margin. Family Nabidae

Fig. 322   Dorsal view of a cotton stainer. Family Pyrrhocoridae

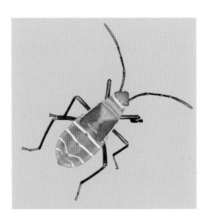

Fig. 323   Immature red bug. Family Pyrrhocoridae

Fig. 324   Venation in the wing of a cotton stainer. Note the long radiating veins. Family Pyrrhocoridae

Suborder Gymnocerata
Long-Horned Bugs

Family Anthocoridae - *Minute Pirate Bugs*   (Fig. 325)
Small (3-5mm).
Antennae with four segments.
Ocelli present.
Cuneus present on corium.
Membrane with few to no veins.

Family Alydidae - *Broad-Headed Bugs*   (Figs. 326-327)
Antennae with four segments and slightly clubbed.
Ocelli present.
Outer width of the head at the eyes is approximately equal to the width of the pronotum.
Membrane with 1-2 large closed cells at the base and many parallel branching veins.

Family Gerridae - *Water Striders*   (Figs. 328-330)
Antennae with four segments.
Ocelli absent.
Legs long and slender.
Middle leg originates closer to the hind leg than to the front leg.
Wings present or absent.
Body pubescent.

Family Veliidae - *Broad-Shouldered Water Striders*   (Fig. 331)
Small (5mm or less).
Antennae with four segments.
Ocelli absent.
Proximal half of body wider than the tapering distal half.
Front tarsi cleft with claws arising before the tips.

Family Hydrometridae - *Water Measurers*   (Figs. 332-333)
Small (8mm or less).
Antennae with four segments.
Ocelli absent.
Wings usually absent.
Long slender body and legs.
Long slender head, equal in length to the thorax.
Bulging eyes arise from sides of head.

Fig. 325   Minute pirate bug.
Family Anthocoridae
Courtesy ARS/USDA

Fig. 326   Dorsal view of a broad-headed bug.
Family Alydidae

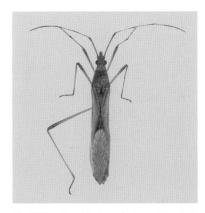

Fig. 327   Dorsal view of a broad-headed bug.
Family Alydidae

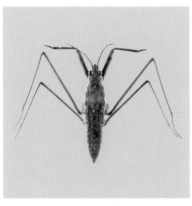

Fig. 328   Dorsal view of a water strider.
Family Gerridae

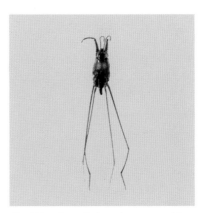

Fig. 329   Dorsal view of a water strider.
Family Gerridae

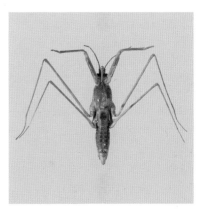

Fig. 330   Ventral view of a water strider showing the leg arrangement.
Family Gerridae

Fig. 331   Dorsal view of a broad-shouldered water strider.
Family Veliidae

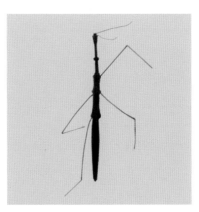

Fig. 332   Dorsal view of a water measurer.
Family Hydrometridae

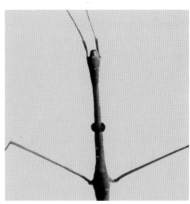

Fig. 333   The long slender head of a water measurer showing the bulging eyes on the side.
Family Hydrometridae

Key to the Families of Hemiptera

1A   Antennae short (shorter than head) and usually concealed in grooves.
                                                          (Suborder Cryptocerata)   **2**

1B   Antennae long (longer than head) and easily visible. **(Suborder Gymnocerata)** **6**

2A   Two long (equal to or greater than half the length of the abdomen) slender, antennae-like structures extending from the tip of abdomen. .... **Nepidae**

2B   Tip of abdomen without long, antennae-like structures. **3**

3A   Medium-sized to large (25-50mm). Body oval and flattened. Two short (less than length of thorax) retractible tubes at the tip of the abdomen. .... **Belostomatidae**

3B   Small to medium-sized (5-15mm). Body variable. No pair of short tubes present at tip of abdomen. **4**

4A   Front femora greatly thickened, almost triangular. Hind legs not flattened and fringed with hair. .... **Naucoridae**

4B   Front femora not enlarged or thickened. Hind legs flattened and fringed with hair. **5**

5A   Body flattened. Dorsum with dark lateral markings. .... **Corixidae**

5B   Body convex or keeled. Dorsum without dark lateral markings. .... **Notonectidae**

6A   Antennae with five segments. **7**

6B   Antennae with four segments. **10**

7A   Small (7mm or less). **8**

7B   Size variable, but 10mm or greater. **9**

8A   Large triangular scutellum present. Tibiae enlarged and densely covered with spines.
                                                                                           Cydnidae

8B   Entire dorsum is covered by scutellum (which appears to be one seamless structure), so that the wings are concealed. Tibiae not enlarged, and spines are weak and not dense when present. .... **Thyreocoridae**

9A   Large triangular scutellum present. Dorsum relatively flat. Body shield-shaped to broadly oval. .... **Pentatomidae**

9B   Entire dorsum is covered by scutellum (which appears to be one seamless structure), so that the wings are concealed. Dorsum convex and rounded. Body shield-shaped to broadly oval or rounded. .... **Scutelleridae**

10A   Ocelli absent. **11**
10B   Ocelli present. **17**

11A   Minute to small (5mm or less). **12**
11B   Size variable, but greater than 5mm. **14**

12A   Body partially to completely covered with lacy, reticulate, sometimes bubbly appearing sculpturing. .... **Tingidae**
12B   Body without lacy, reticulate, bubbly appearing sculpturing. **13**

13A   Body flattened, round to oval in shape. Wingless. .... **Cimicidae**
13B   Body not flattened. Proximal half is wider than distal half which tapers towards the tip. Wings may be absent or present. .... **Veliidae**

14A   Long slender body and legs. Long (equal to the length of the thorax) slender head with bulging eyes that arise from the sides. .... **Hydrometridae**
14B   Body and legs not long and slender. Head not long and slender with bulging eyes. **15**

15A   Body extremely flattened. Wings cover only central portion of dorsum with abdomen extending beyond their edges. .... **Aradidae**
15B   Body not extremely flattened. Wings cover almost entire abdomen. **16**

16A   Cuneus present. Wingtip tilts down when viewed in profile. .... **Miridae**
16B   Cuneus absent. Wingtip does not tilt down when viewed in profile. .... **Pyrrhocoridae**

17A   Minute to small (3-5mm). Cuneus present. Black and white markings. ...... **Anthocoridae**
17B   Size variable, but greater than 5mm. Cuneus absent. Color and markings variable. **18**

18A   Abdomen not completely covered by wings, extending beyond their edges when viewed dorsally. Front legs often raptorial. **19**
18B   Abdomen completely covered by wings, or elongate if wings absent. **20**

19A   Front femora enlarged and extremely thickened. Antennae slightly clubbed. Tip of beak does not fit in a groove between the front legs. .... **Phymatidae**
19B   Front legs may be slightly enlarged, but not extremely thickened. Antennae not clubbed. Tip of beak fits into a groove between the front legs. .... **Reduviidae**

20A   Body slender and elongate. Middle legs arise noticeably closer to the hind legs than to the front legs. Hind femur extends beyond the tip of the abdomen (when positioned along side of it). .... **Gerridae**

20B   Body variable. All pairs of legs are approximately equidistant from one another. Hind femur does not extend beyond the tip of the abdomen. **21**

21A   Width of head at eyes equal to width of pronotum. Membrane with one to two large closed cells at base and many parallel and branching veins beyond. .... **Alydidae**

21B   Width of head at eyes less than width of widest part of pronotum. Membrane variable. **22**

22A   Small to medium-sized (5-20mm). Membrane with 4-5 parallel veins. Hind tibiae not flattened and expanded, and without leaf-like flaps. .... **Lygaeidae**

22B   Medium-sized to large (15-40mm). Membrane with many (10-20) parallel veins. Hind tibiae often flattened and expanded or with leaf-like flaps. .... **Coreidae**

This tropical leaf-footed bug was found in the mountain forests of Trinidad. The bright patches on the hind legs serve to advertise to predators the unpleasant taste of this insect. Many tropical leaf-footed bugs with warning coloration feed on passion vines from which they derive the chemicals that make their tissues unpalatable. Although some U.S. species of the family Coreidae have similar flaps on the hind legs, none are so vividly colored as those found in the tropics.

Order Thysanoptera
Thrips

Characteristics of Thysanoptera

Minute (4mm or less).
Body torpedo-shaped.
Wings present or absent.
Four long slender membranous wings with an extremely long fringe of hairs around the edges.
Antennae short (less than the length of the head and thorax combined).
Mouthparts rasping-sucking and asymmetrical.
Metamorphosis intermediate between gradual and complete.

Suborder Tubulifera - *Tube-Tailed Thrips*   (Figs. 334-335)

Tip of abdomen tubular.
No ovipositor present in females.
Forewings without veins and setae.

Suborder Terebrantia - *Round-Tailed Thrips*   (Figs. 336-337)

Tip of the abdomen rounded or conical.
Ovipositor present and visible in females.
Forewings with veins and setae.

Fig. 334   The Cuban laurel thrips is an example of a tube-tailed thrips.
Suborder Tubulifera

Fig. 335   Photomicrograph of a tube-tailed thrips. Note the pointed tip of the abdomen.   Suborder Tubulifera

Fig. 336   A flower thrips is an example of a round-tailed thrips. Suborder Terebrantia
Photo JL Castner   Courtesy IFAS/UF

Fig. 337   Photomicrograph of a round-tailed thrips. Note the rounded tip of the abdomen.   Suborder Terebrantia

Order Homoptera
Cicadas, Hoppers, Aphids, and Others

Characteristics of Homoptera

Presence and number of wings variable.
Mouthparts consist of a piercing-sucking 'beak', with no palps.
Beak arises from the posterior or back portion of the underside of the head.
Antennae variable.
Ocelli present or absent.
Gradual metamorphosis.

Suborder Auchenorrhyncha
Cicadas and Hoppers

Characteristics of Auchenorrhyncha

Two pairs of wings, held roof-like over the body.
Wings membranous or only slightly thickened and leathery.
Antennae short (much less than the length of the body).
Tarsi with three segments.

Family Cicadidae - *Cicadas*   (Fig. 338)

Large insects (25-50mm).
Wings membranous.
Three ocelli present.
Prominent eyes.
Distinctive shape.

Fig. 338   Cicadas vary in size but have the typical body shape seen here. Family Cicadidae

Family Cicadellidae - *Leafhoppers*
(Figs. 339-340)

Small-medium (20mm or less).
Resemble small cicadas in general form.
Front wings usually opaque, pigmented, and slightly leathery.
Body parellel-sided or tapers posteriorly.
Prominent eyes.
Hind tibiae with one or more rows of small spines.

Fig. 339   Tropical leafhopper. Note the row of spines along the hind tibia. Family Cicadellidae

Suborder Auchennorhyncha
Cicadas and Hoppers

Family Cercopidae - *Froghoppers*　　(Figs. 341-342)
Small (12mm or less).
Hind tibiae with 1-2 strong spines along the edge and a ring of spines at the distal tips.
Forewings slightly leathery.
Widest part of body usually midway between the head and tail.

Family Membracidae - *Treehoppers*　　(Figs. 343-345)
Small (12mm or less)
Pronotum projects backwards over the abdomen, and sometimes forwards over the head.
Some species with very unusual shapes.

Family Dictyopharidae - *Dictyopharid Planthoppers*
Antenna originates on the side of the head beneath the eye.
Some species with head extended forward into a snout-like projection.　　(Figs. 346-347)
Some species with 2-3 raised ridges on the head, and the front femora flattened and dilated.

Family Fulgoridae - *Fulgorid Planthoppers*　　(Fig. 348)
Medium-large (reaching 25mm).
Antenna originates on the side of the head beneath the eye.
Many cross veins in anal area of wing.
Ocelli (when present) on side of head in front of eyes.

Family Acanaloniidae - *Acanaloniid Planthoppers*　　(Figs. 349-350)
Antenna originates on the side of head beneath the eye.
Wings normally held almost vertically.
Wings broadly oval to almost semicircular when viewed laterally.
Costal and apical margins of forewings with reticulate pattern of many small rounded cells.

Family Flatidae - *Flatid Planthoppers*　　(Fig. 351)
Antenna originates on the side of the head beneath the eye.
Wings normally held almost vertically.
Wings wedge-shaped when viewed laterally.
Costal and apical margins of forewings with pattern of many small squarish to rectangular-shaped cells.

**Fig. 340   A 'sharpshooter' planthopper.
Family Cicadellidae**

**Fig. 341   Froghopper, the adult stage of
a 'spittlebug'.
Family Cercopidae**

**Fig. 342   Tropical froghopper.
Family Cercopidae**

**Fig. 343   Thorn 'bug', a treehopper with
a spine-like pronotum.
Family Membracidae**

**Fig. 344   Treehopper.
Family Membracidae**

**Fig. 345   Alfalfa treehopper.
Family Membracidae**

**Fig. 346   Lateral view of a dictyopharid
planthopper.
Family Dictyopharidae**

**Fig. 347   Dorsal view of a dictyopharid
planthopper.
Family Dictyopharidae**

**Fig. 348   Tropical fulgorid planthopper.
Family Fulgoridae**

**Fig. 349   Acanaloniid planthopper.
Family Acanaloniidae**

**Fig. 350   Tropical acanaloniid
planthopper.
Family Acanaloniidae**

**Fig. 351   Flatid planthopper.
Family Flatidae**

Suborder Sternorrhyncha
Aphids, Whiteflies, and Scales

Characteristics of Sternorrhyncha

Antennae long and thread-like.
Tarsi with one or two segments.
Some families with distinctive-shaped waxy coverings over the body.

Family Aphididae - *Aphids or Plantlice*   (Figs. 352-354)

Small (8mm or less).
Wings present or absent.
Pair of dorsal tubes (cornicles) project from posterior part of the body.
Body pear-shaped and soft.

Family Aleyrodidae - *Whiteflies*   (Fig. 355)

Minute (3mm or less).
Body and wings covered with a whitish dust.
Forewings and hindwings nearly the same size, and held roof-like over the body.

Superfamily Coccoidea: Most of this group secrete a waxy covering that completely hides
the body. Only adult females, which are wingless and usually legless, will be discussed here. Adult
males have one pair of wings, a pair of abdominal filaments, and sometimes a terminal style or filament.

Family Diaspididae - *Armored Scales*   (Figs. 356-357)

Small (4mm or less).
Waxy secretion is hard.
Slightly convex when viewed laterally.
Body shape usually oval, rounded, or like an oyster shell.

Family Coccidae - *Tortoise & Wax Scales*   (Figs. 358-360)

Small (8mm or less).
Waxy secretion is soft.
Markedly convex when viewed laterally.
Body shape variable.

Family Pseudococcidae - *Mealybugs*   (Fig. 361)

Small (10mm or less).
Body soft with a white waxy coating.
Oval-shaped with a marginal fringe of projections.

Family Margarodidae - *Ground Pearls & Giant Coccids*

Small-medium in size (20mm or less).

(Figs. 362-363)

Generally rounded in shape.
Some species form metallic-appearing wax cysts or produce large amounts of waxy plates and tendrils.

Fig. 352   Cluster of turnip aphids.
Family Aphididae

Fig. 353   Wingless aphid with large, distinctive cornicles.
Family Aphididae

Fig. 354   Winged milkweed aphid.
Family Aphididae

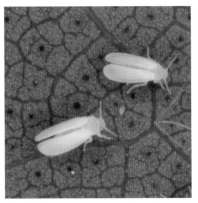

Fig. 355   Adult sweetpotato whiteflies.
Family Aleyrodidae

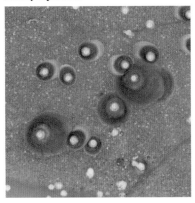

Fig. 356   Florida red scale, one of the armored scales.
Family Diaspididae

Fig. 357   White magnolia scale. Note the hard brown waxy secretions of this armored scale.   Family Diaspididae

Fig. 358   Hemispherical scales. Note the convex shape of the hard covering.
Family Coccidae

Fig. 359   Florida wax scale.
Family Coccidae

Fig. 360   Tuliptree scale, one of the tortoise scales.
Family Coccidae

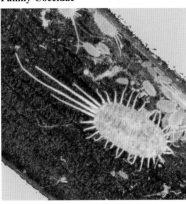

Fig. 361   Long-tailed mealybugs.
Family Pseudococcidae

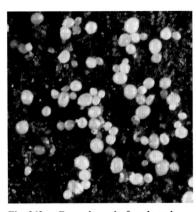

Fig. 362   Ground pearls, female scale insects encased by a round wax cyst.
Family Margarodidae
Photo JL Castner   Courtesy IFAS/UF

Fig. 363   Cottony cushion scale.
Family Margarodidae

Key to the Families of Homoptera

1A   Antennae short (usually less than length of head and thorax combined), and bristle-like. Tarsi with three segments. **(Suborder Auchenorrhyncha)** **2**

1B   Antennae long (usually greater than length of head and thorax combined), and thread-like. Tarsi with one or two segments. **(Suborder Sternorrhyncha)** **9**

1C   Appearance not insect-like, with legs absent or not visible. A waxy covering (variable in shape, size, hardness, and thickness) present.

                                   (Suborder Sternorrhyncha)(Superfamily Coccoidea) **10**

2A   Antennae originate between or in front of eyes on front of head. **3**

2B   Antennae originate beneath eyes on sides of head. **6**

3A   Pronotum greatly extended over abdomen and head, sometimes forming unusual shapes.

                                                     Membracidae

3B   Pronotum does not extend over abdomen or head. **4**

4A   Medium-sized to large (25-50mm). Three ocelli present. .... **Cicadidae**

4B   Small to medium-sized (20mm or less). Two ocelli or none present. **5**

5A   Hind tibia with one or more rows of small spines along its length. Body either parallel-sided or tapers posteriorly when viewed dorsally. .... **Cicadellidae**

5B   Hind tibia with one or two spines along its length and a circle of spines at its apex. Body wider at middle when viewed dorsally. .... **Cercopidae**

6A   Wings held almost flat together, vertically above and around the body. **7**

6B   Wings held roof-like over and around the body. **8**

7A   Costal and apical margins of forewings with pattern of small squarish to rectangular-shaped cells. .... **Flatidae**

7B   Costal and apical margins of forewings with reticulate pattern of many small, rounded cells. .... **Acanaloniidae**

8A   Head extended forward into a beak-like snout or bearing 2-3 ridges in front. Anal area of hindwing without many cross veins. .... **Dictyopharidae**

8B   Head not extended forward and without ridges. Anal area of hindwing with many cross veins. .... **Fulgoridae**

9A   Small (4-8mm), soft-bodied and pear-shaped. A pair of tube-like structures on dorsum near tip of abdomen. Wings clear when present. .... **Aphididae**

9B   Minute (2-3mm), with body and wings covered by a whitish powder. No pair of tube-like structures near tip of abdomen. Wings white and always present. .... **Aleyrodidae**

10A   Body gray or white, flattened, and oval-shaped with a marginal fringe of filamentous projections. .... **Pseudococcidae**

10B   Body completely concealed by waxy covering. Characteristics of covering are variable.   **11**

11A   Waxy covering in the form of hard, rounded cyst or waxy plates and tendrils. Small to medium-sized (10-20mm). .... **Margarodidae**

11B   Waxy covering variable, but not as above. Small (10mm or less). **12**

12A   Waxy covering is soft, but thick and convex when viewed laterally. .... **Coccidae**

12B   Waxy covering is hard and usually flattened and disc-like. .... **Diaspididae**

Order Mecoptera
Scorpionflies

Characteristics of Mecoptera

Small-medium (5-25mm).
Head extended downwards giving the face a 'long', horse-like appearance.
Tip of abdomen may be swollen and bulbous resembling a scorpion's sting.
Four wings usually present, membranous and approximately the same size.
Wings pigmented or plain.
Antennae thread-like.
Tarsi with five segments.
Chewing mouthparts (located at the tip of the face).
Complete metamorphosis.

Family Bittacidae - *Hangingflies*

Medium-sized (15-25mm).
Tip of male abdomen enlarged, but does not look like a scorpion's sting.
Tarsi with one claw.
Wings narrow at the base, almost stalk-like as in some damselflies.

Fig. 364   Dorsal view of a hanging scorpionfly.   Family Bittacidae

Family Boreidae - *Snow Scorpionflies*

Small (4-5mm).
Wings absent, except for small vestiges.
Tarsi with two claws.

Fig. 365   Photomicrograph of a snow scorpionfly.   Family Boreidae

Family Panorpidae - *Common Scorpionflies*

Medium-sized (15-20mm).
Wings often pigmented.
Tip of the male abdomen resembles a scorpion's sting.
Tarsi with two claws.

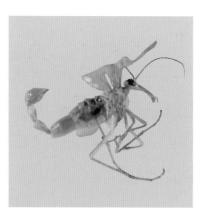

Fig. 366   Lateral view of a common scorpionfly.
Family Panorpidae

Fig. 367   Extended head of a common scorpionfly.
Family Panorpidae

Fig. 368   Lateral view of a common scorpionfly.
Family Panorpidae

Order Neuroptera
Lacewings, Antlions, and Others

Characteristics of Neuroptera

Two pairs of large, membranous wings usually held roof-like over the body.
Wings with both many longitudinal veins and many cross veins.
Antennae long and variable.
Ocelli present or absent.
Cerci absent.
Chewing mouthparts.
Complete metamorphosis.

Suborder Raphidiodea
Snakeflies

Characteristics of Raphidiodea

Prothorax long (usually at least one half the length of the abdomen).
Front legs originate from the posterior part of the thorax.
Front legs similar to the other legs, not raptorial.
Female with long (greater than half the length of the abdomen) ovipositor.

Family Raphidiidae   (*Raphidiid Snakeflies*)

Small (5-15mm).
Ocelli present.
Cross vein present at the proximal edge of the stigma in the front wing.

Family Inoceliidae   (*Inoceliid Snakeflies*)

Small-medium (10-20mm).
Ocelli absent.
Proximal edge of the stigma of the front wing is not bordered by a cross vein.

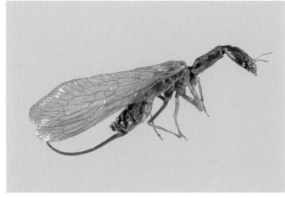

Fig. 369　Raphidiid snakefly.
Family Raphidiidae

Fig. 370　Inoceliid snakefly.
Family Inoceliidae

Suborder Planipennia
Lacewings, Mantispids, Antlions, and Owlflies

Characteristics of the Planipennia

Forewings and hindwings similar in size and shape.
Prothorax not lengthened, except in the Family Mantispidae.

Family Mantispidae - *Mantispids or Mantidflies*   (Figs. 371-373)

Medium-sized (20-30mm).
Front legs raptorial and with spines.
Head triangular and mantid-like, with large eyes.
Prothorax lengthened and extended.
Front legs originate from the anterior part of the thorax.
Some species resemble wasps.

Family Myrmeleontidae - *Antlions*   (Figs. 374-376)

Medium-large (40-80mm).
Antennae knobbed, approximately as long as head and thorax combined.
Elongate closed cell is present in wingtip just below the subcosta.
Base of male hindwing often with small knobbed organ.

Family Ascalaphidae - *Owlflies*   (Figs. 377-379)

Medium to large (40-50mm).
Antennae knobbed and long (approximately as long or longer than the body).
Eyes large and bulging, sometimes divided by a groove.

Family Chrysopidae - *Green Lacewings*   (Figs. 380-381)

Small-medium-sized (35mm or less).
Antennae long (usually as long or longer than the body) and thread-like.
Cross veins between the front edge (costa) of the wing and the next
longitudinal vein that parallels it (subcosta) are not forked.
Body and wings usually greenish and not hairy.
Eyes usually gold or copper in color.
Most species with a tympanum at the base of the forewing.

Family Hemerobiidae - *Brown Lacewings*   (Fig. 382)

Small-medium (10-20mm).
Antennae long (as long as the body or longer) and thread-like.
Some cross veins between the costa and subcosta (see Chrysopidae) are forked.
Body and wings usually brown and hairy.

**Fig. 371   Mantidfly that resembles a green lacewing.
Family Mantispidae**

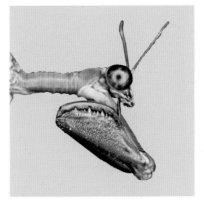

**Fig. 372   Raptorial forelegs of a mantidfly.
Family Mantispidae**

**Fig. 373   Mantidfly that resembles a paper wasp.
Family Mantispidae**

Fig. 374   Antlion adult. Note the long membranous wings similar to those of a damselfly.   Family Myrmeleontidae

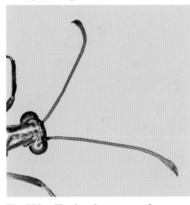

**Fig. 375   Head and antennae of an antlion.
Family Myrmeleontidae**

**Fig. 376   Antlion in typical resting posture.
Family Myrmeleontidae**

**Fig. 377   Dorsal view of an owlfly with the wings spread.
Family Ascalaphidae**

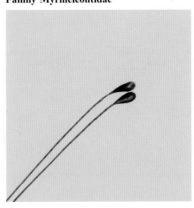

**Fig. 378   Distal portion of owlfly antennae showing the terminal knobs.
Family Ascalaphidae**

**Fig. 379   Owlfly in typical resting posture.
Family Ascalaphidae**

**Fig. 380   Dorsal view of a green lacewing.
Family Chrysopidae**

**Fig. 381   Green lacewing.
Family Chrysopidae**

**Fig. 382   Brown lacewing.
Family Hemerobiidae**
Photo JL Castner   Courtesy IFAS/UF

Suborder Megaloptera
Dobsonflies, Fishflies, and Alderflies

Characteristics of Megaloptera

Hindwings wider than the forewings at the base.
Anal area of the hindwings folds like a fan when at rest.

Family Corydalidae   (*Dobsonflies & Fishflies*)

Medium-large (25-80mm).
Ocelli present.
Fourth tarsal segment cylindrical.
Antennae variable.
Males sometimes with very long mandibles (greater than 2X the length of the head).

Family Sialidae   (*Alderflies*)

Small-medium (less than 25mm).
Ocelli absent.
Fourth tarsal segment widened and has two lobes.

Fig. 383   Dorsal view of a male dobsonfly. Note the large mandibles.
Family Corydalidae.

Fig. 384   The head of a dobsonfly. Note the ocelli.
Family Corydalidae

Fig. 385   Dorsal view of a female dobsonfly with the wings spread.
Family Corydalidae

Fig. 386   Dorsal view of an alderfly.
Family Sialidae

Order Coleoptera
Beetles

Characteristics of Coleoptera

Two pairs of wings usually present.
Forewings modified into hardened protective covers (elytra) that meet in a straight line down the back.
Hindwings are entirely membranous, usually folded and hidden beneath the elytra.
Chewing mouthparts.
Complete metamorphosis.
Extremely diverse order with many families.

Terminology

elytron - Forewing that is thickened and hard, concealing the membranous hindwing beneath it.
scutellum - Often triangular-shaped structure located on the dorsum between the bases of the wings.
tarsal formula - Reference to the number of tarsal segments found on the pro-, meso-, and metathoracic legs, respectively. An example would be the Family Tenebrionidae with the tarsal formula 5-5-4.

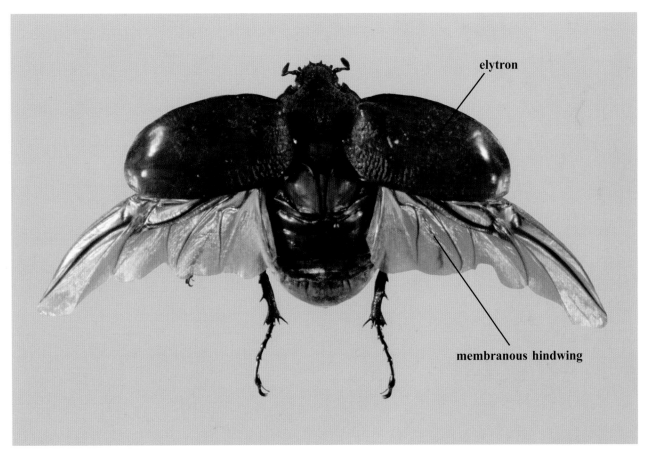

Fig. 387   Adult beetle with the wings spread to show the difference in the hardened forewings (elytra) and the membranous hindwings that are used for flight.

Suborder Adephaga
Adephagous Beetles

Characteristics of Adephaga
First abdominal segment is completely divided ventrally
by the hind coxae (Fig. 388).
Antennae usually thread-like.
Tarsal formula usually 5-5-5.
Trochanters large (2-3X size of coxae).

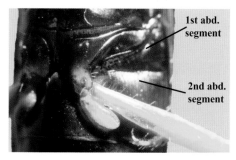

**Fig. 388   Ventral view of an adephagous beetle
showing how the hind coxae divide the first
abdominal segment.**

Family Cicindelidae - *Tiger Beetles*   (Figs. 389-391)
Eyes large and bulbous.
Width of head at eyes as wide or wider than the pronotum.
Large toothed mandibles.
Body shape distinctive.
Elytra often patterned, but usually without ridges or rows of punctations.
Antennae originate above the base of the mandibles.

Family Carabidae - *Ground Beetles*   (Figs. 392-394)
Eyes often large and bulbous.
Width of head at eyes narrower than the pronotum.
Elytra seldom patterned, but typically with ridges and rows of punctations.
Body shape extremely variable.
Antennae originate between the base of the mandibles and the eyes.

Family Dytiscidae - *Predaceous Diving Beetles*   (Figs. 395-397)
Hind legs long, flattened, and fringed with hairs.
Antennae long and thread-like.
Body convex and oval in shape.
Scutellum small, but usually visible.

Family Gyrinidae - *Whirlygig Beetles*   (Figs. 398-400)
Two pairs of compound eyes (one above the other).
Hind legs and middle legs very short, flattened, and not fringed with hairs.
Antennae short and clubbed.
Body flattened and oval in shape.
Scutellum not visible.

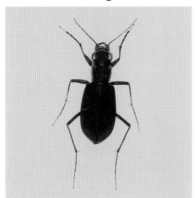

Fig. 389 Dorsal view of a tiger beetle. Note the distinctive shape. Family Cicindelidae

Fig. 390 Head and thorax of a tiger beetle. Note the eyes are wider than the pronotum. Family Cicindelidae

Fig. 391 Large toothed mandibles and bulbous eyes of a tiger beetle. Family Cicindelidae

Fig. 392 Dorsal view of a ground beetle. The shape of this group may vary. Family Carabidae

Fig. 393 Head and thorax of a ground beetle. Note the pronotum is wider than the eyes. Family Carabidae

Fig. 394 Body shape in ground beetles varies greatly, as seen in this specimen. Family Carabidae

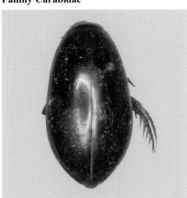

Fig. 395 Dorsal view of a predaceous diving beetle, which has a convex and oval body. Family Dytiscidae

Fig. 396 The hind leg of a predaceous diving beetle is flattened and fringed with long hairs. Family Dytiscidae

Fig. 397 The head and antennae of a predaceous diving beetle. Family Dytiscidae

Fig. 398 Dorsal view of a whirlygig beetle showing the oval, flattened body. Family Gyrinidae

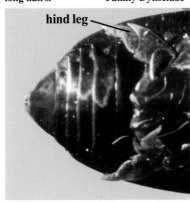

Fig. 399 The hind legs of a whirlygig beetle are short and flattened, but not fringed with hairs. Family Gyrinidae

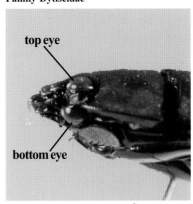

Fig. 400 Lateral view of the head of a whirlygig beetle showing the double compound eye on each side. Family Gyrinidae

Suborder Polyphaga
Polyphagous Beetles

Characteristics of Polyphaga

First abdominal segment not completely divided ventrally
by the hind coxae (Fig. 401).
Trochanters small (equal to or less than the size
of the hind coxae).
Antennae variable.
Tarsal formulae variable.

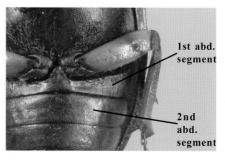

Fig. 401   Ventral view of the abdomen of a polyphagous beetle showing how the hind coxae do not divide the first abdominal segment.

Family Hydrophilidae - *Water Scavenger Beetles* (Figs. 402-404)

Hind legs flattened and fringed with hairs.
Backward-projecting spine between legs on venter of thorax.
Antennae short and clubbed, sometimes hidden.
Palps long (may be mistaken for antennae).
Body oval and convex.

Family Histeridae - *Clown Beetles*   (Figs. 405-407)

Small (10mm or less).
Elytra short, exposing the tip of the abdomen (1-3 segments) dorsally.
Antennae short, elbowed, and clubbed.
Tibiae usually wide and flattened.
Body often black and shiny.

Family Staphylinidae - *Rove Beetles*   (Figs. 408-410)

Shape distinctive: elongate, slender, and parallel-sided.
Elytra short (equal to or only slightly longer than the pronotum in length).
Much of the abdomen usually exposed (1-6 segments) dorsally.
Body appears to be divided into four regions when viewed dorsally.
Antennae thread-like or clubbed.
Size variable, although shape remains fairly consistent.

Family Silphidae - *Carrion & Sexton Beetles*   (Figs. 411-413)

Elytra slightly to moderately short, exposing the tip of the abdomen (1-3 segments) dorsally.
Elytra usually flattened and leathery rather than hard and horny. Often only loosely cover body.
Elytra tend to widen posteriorly.
Antennae clubbed.
Tarsal formula 5-5-5.

fig. 402   Dorsal view of a water scavenger beetle showing the oval convex body.   Family Hydrophilidae

Fig. 403   Ventral view of a water scavenger beetle showing the large spine between the legs.   Family Hydrophilidae

spine

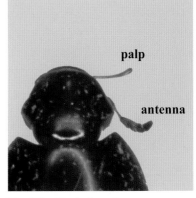

palp

antenna

Fig. 404   Head of a water scavenger beetle showing the slender palp and clubbed antenna.   Family Hydrophilidae

Fig. 405   Clown or hister beetle. Note the short elytra.
Family Histeridae

Fig. 406   Clown or hister beetle. Note the typical convex body shape.
Family Histeridae

Fig. 407   Dorsal view of an extremely flattened species of clown beetle.
Family Histeridae

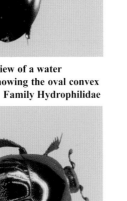

pronotum

elytron

Fig. 408   Dorsal view of a rove beetle showing the typical and distinctive body shape.   Family Staphylinidae

Fig. 409   A common rove beetle that is attracted to carrion.
Family Staphylinidae

Fig. 410   The hairy rove beetle, a species that is attracted to carrion.
Family Staphylinidae

Fig. 411   Many carrion beetles have orange and black markings such as this.
Family Silphidae

Fig. 412   A carrion beetle. Note that the tip of the abdomen is exposed.
Family Silphidae

Fig. 413   The American carrion beetle. Note that the tip of the abdomen is exposed.   Family Silphidae

Suborder Polyphaga
Polyphagous Beetles

Family Lampyridae - *Lightningbugs or Fireflies* (Figs. 414-416)
Body elongate and parallel-sided.
Head concealed by pronotum when viewed dorsally.
Tip of abdomen (when viewed ventrally) lighter in color than the rest of the abdomen.
Elytra leathery rather than hard and shell-like.
Antennae thread-like or sawtoothed.
Fourth tarsal segment modified into flattened lobes.
Tarsal formula 5-5-5.

Family Cantharidae - *Soldier Beetles*   (Figs. 417-419)
Body elongate and parallel-sided.
Head normally not concealed by pronotum when viewed dorsally.
Tip of abdomen (when viewed ventrally) same color as the rest of the abdomen.
Elytra leathery rather than hard and shell-like.
Antennae usually thread-like.
Fourth tarsal segment modified into flattened lobes.
Tarsal formula 5-5-5.

Family Lycidae - *Net-Winged Beetles*   (Figs. 420-422)
Elytra leathery rather than hard and shell-like.
Elytra with pronounced longitudinal ridges that have a reticulate network of veins between them.
Elytra become wider posteriorly, not parallel-sided.
Elytra usually colored with red, black, orange, or a combination.
Majority of the head concealed by the pronotum when viewed dorsally.
Antennae flattened, either thread-like or sawtoothed.

Family Dermestidae - *Skin or Carpet Beetles*   (Figs. 423-425)
Small (12mm or less).
Body oval to round in shape.
Body often covered with hairs or scales.
Antennae clubbed and fit into grooves.
Median ocellus may be present.
Tarsal formula 5-5-5.

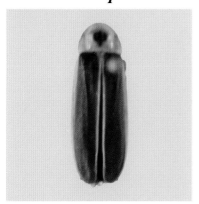

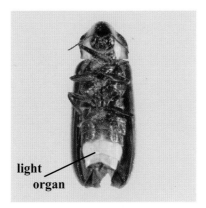

Fig. 414   Dorsal view of a lightningbug or firefly. Note that the pronotum conceals the head.   Family Lampyridae

Fig. 415   Ventral view of a firefly showing the light-producing organ. Family Lampyridae

Fig. 416   Head and thorax of a firefly. Note the bulging eyes. Family Lampyridae

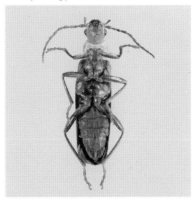

Fig. 417   Dorsal view of a soldier beetle. Note that the pronotum does not conceal the head.   Family Cantharidae

Fig. 418   Ventral view of a soldier beetle. Note the absence of any light-producing organ.   Family Cantharidae

Fig. 419   Head and thorax of a soldier beetle. Family Cantharidae

Fig. 420   Dorsal view of a net-winged beetle. Note how the elytra widen towards the tip.   Family Lycidae

Fig. 421   Many net-winged beetles are brightly colored such as this one. Family Lycidae

Fig. 422   Dorsal view of a net-winged beetle. Family Lycidae

Fig. 423   Oblique view of a skin beetle or dermestid beetle. Family Dermestidae

Fig. 424   Varied carpet beetles can be pests in the home. Family Dermestidae

Fig. 425   Skin beetles such as this one are sometimes found on dried and/or decomposing animal tissues. Family Dermestidae

Suborder Polyphaga
Polyphagous Beetles

Family Elateridae - *Click Beetles*   (Figs. 426-428)
Body elongate with elytra narrowing posteriorly.
Posterior corners of pronotum pointed.
Backward-pointing spine on venter, originating between the front pair of legs
and fitting into a groove between the middle pair of legs.
Antennae usually sawtoothed.
Tarsal formula 5-5-5.

Family Scarabaeidae - *Scarab Beetles*   (Figs. 429-431)
Robust, heavy-bodied, convex beetles.
Antennae lamellate, with last 3-7 segments composed of flattened
lobes that can be held tightly together.
Front tibiae widened with the outer edges toothed.
Size and color variable.
Tarsal formula 5-5-5.

Family Passalidae - *Bessbugs or Betsy Beetles*   (Figs. 432-434)
Large (30-40mm), and shiny black or brown in color.
Body wide and parallel-sided.
Pronotum distinctly separated from the elytra.
Pronotum with a single median longitudinal groove.
Head with a small median horn.
Last three segments of antennae form a club, but cannot be held tightly together.
Tarsal formula 5-5-5.

Family Lucanidae - *Stag Beetles*   (Figs. 435-437)
Size variable (10-60mm).
Black to red-brown in color.
Body wide and parallel-sided.
Pronotum separated from the elytra.
Pronotum without a median longitudinal groove.
Antennae elbowed, with last 3-4 segments forming a club that cannot be held tightly together.
Mandibles may be very large.
Tarsal formula 5-5-5.

Fig. 426   Click beetle. Note pointed corners of pronotum. Family Elateridae

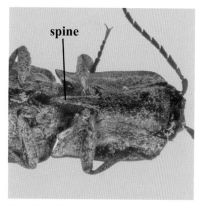

spine

Fig. 427   Ventral view of a click beetle showing the spine between the legs. Family Elateridae

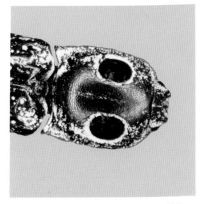

Fig. 428   The eyed elater, a large click beetle with eyespots on the pronotum. Family Elateridae

Fig. 429   Scarab beetle. Family Scarabaeidae

Fig. 430   The head and antennae of a scarab beetle. Note how the antennal lobes are spread.   Family Scarabaeidae

Fig. 431   Head, antenna, and front leg of a scarab beetle. Note how the antennal lobes are closed.   Family Scarabaeidae

Fig. 432   Bessbug or betsy beetle. Family Passalidae

Fig. 433   Head and thorax of a betsy beetle. Note the median groove in the pronotum.   Family Passalidae

Fig. 434   A bessbug or betsy beetle. Note the grooves in the elytra. Family Passalidae

Fig. 435   Dorsal view of a stag beetle. Family Lucanidae

Fig. 436   Head and antenna of a stag beetle. Family Lucanidae

Fig. 437   Head and mandibles of a stag beetle. Family Lucanidae

Suborder Polyphaga
Polyphagous Beetles

Family Buprestidae - *Metallic Wood-Boring Beetles*

Body with characteristic 'bullet' shape.
Cuticle with a metallic sheen (especially ventrally).   **(Figs. 438-440)**
Antennae sawtoothed or thread-like.
Elytra come together posteriorly in a blunt point.
Size and color variable, although shape is fairly consistent.
Tarsal formula 5-5-5.

Family Coccinellidae - *Ladybird Beetles or Ladybugs*

Small (10mm or less).
Body broadly oval to round, and extremely convex.   **(Figs. 441-443)**
Usually brightly colored or spotted with red, orange, or black.
Head partially to entirely concealed by pronotum when viewed dorsally.
Antennae short, with last 3-6 segments forming a club.
Tarsi with very small third segment, giving the tarsal formula the
appearance of 3-3-3 when it is actually 4-4-4.

Family Meloidae - *Blister Beetles*   (Figs. 444-446)

Distinctive body shape with pronotum narrower than both the head and the base of the elytra.
Elytra soft and leathery, curving loosely over the abdomen, sometimes exposing the tip.
Antennae thread-like or bead-like.
Color variable.
Tarsal formula 5-5-4.

Family Tenebrionidae - *Darkling Beetles*   (Figs. 447-449)

Margin of eye notched by a keel-like ridge on the head.
Antennae are thread-like, bead-like, or slightly clubbed.
Body shape variable with color usually a dull black, brown, or red.
Tarsal formula 5-5-4.

Fig. 438   Tropical metallic wood-boring beetle shows the typical body shape. Family Buprestidae

Fig. 439   Some metallic wood-boring beetles have cryptic, bark-like colors. Family Buprestidae

Fig. 440   Many buprestid beetles are brightly pigmented or brilliantly metallic. Family Buprestidae

Fig. 441   A ladybug or ladybird beetle. Note the bright colors and convex body. Family Coccinellidae

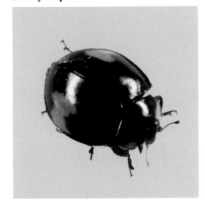

Fig. 442   Twice-stabbed ladybug. Family Coccinellidae

Fig. 443   The Mexican bean beetle is an important economic pest. Family Coccinellidae

Fig. 444   Blister beetle. Family Meloidae

Fig. 445   Head and thorax of a blister beetle. Note the wide head and narrow pronotum.   Family Meloidae

Fig. 446   Blister beetle. Family Meloidae

Fig. 447   Flour beetle, the adult form of a mealworm. Family Tenebrionidae

Fig. 448   Darkling beetle. Family Tenebrionidae

Fig. 449   Lateral view of the forked fungus beetle. Family Tenebrionidae

Suborder Polyphaga
Polyphagous Beetles

Family Cerambycidae - *Long-Horned Beetles*   (Figs. 450-452)

Antennae long (at least half the length of the body, but usually much longer).
Body elongate and cylindrical in shape.
Eyes partially to deeply notched by bases of the antennae.
First antennal segment 4-5X as long as the second antennal segment.
Size variable.
Third tarsal segment bilobed.
Tarsal formula appears to be 4-4-4 but is 5-5-5.

Family Chrysomelidae - *Leaf Beetles*   (Figs. 453-455)

Small-medium (20mm or less).
Antennae less than half the length of the body long.
Body shape variable from rounded to broadly oval to pear-shaped.
Third tarsal segment bilobed.
Tarsal formula appears to be 4-4-4 but is 5-5-5.

Family Curculionidae - *Snout Beetles*   (Figs. 456-458)

Head with a well developed 'snout', ranging from short and broad to long slender and curving.
Antennae elbowed, with the last three segments forming a club.
Antennae arise from the middle of the 'snout'.
Size variable.
Tarsal formula appears to be 4-4-4 but is 5-5-5.

Family Cleridae - *Checkered Beetles*   (Figs. 459-461)

Body elongate, parallel-sided, and covered with hairs.
Head as wide or wider than pronotum.
Pronotum narrower than the bases of the elytra.
Antennae variable.
Body often colorful.
Tarsal formula 5-5-5.

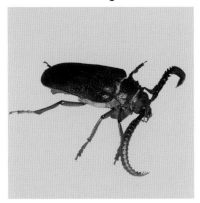

Fig. 450   Long-horned wood-boring beetle. Note the extensive antennae. Family Cerambycidae

Fig. 451   A long-horned wood-boring beetle with fairly short antennae. Family Cerambycidae

Fig. 452   Head of a long-horned wood-boring beetle. Note the notch in the eye. Family Cerambycidae

Fig. 453   The Colorado potato beetle is one of many pest leaf beetles. Family Chrysomelidae

Fig. 454   The spotted cucumber beetle is a destructive leaf beetle species. Family Chrysomelidae

Fig. 455   Tortoise beetle. Note the rounded shape and metallic colors. Family Chrysomelidae

Fig. 456   Large palm weevil. Note the long snout and elbowed antennae. Family Curculionidae

Fig. 457   The plum curculio, a pest weevil. Family Curculionidae

Fig. 458   A species of blunt-nosed weevil. Note the short broad snout. Family Curculionidae

Fig. 459   Checkered beetle. Family Cleridae

Fig. 460   The red-legged ham beetle is often found on carrion and dried animal material.   Family Cleridae

Fig. 461   Predatory checkered beetle. Family Cleridae *Courtesy ARS/USDA*

123

Key to the Families of Coleoptera

1A   Hind coxae large and curving, extending across the entire first abdominal segment beyond the point where it borders the second. (**Suborder Adephaga**) **2**

1B   Hind coxae not large and curving, failing to extend across the entire first abdominal segment. (**Suborder Polyphaga**) **5**

2A   Each compound eye is clearly separated into an upper eye and lower eye. Antennae are short (less than the length of the head and thorax combined) and clubbed. .... **Gyrinidae**

2B   Each compound eye is not divided. Antennae are long (equal to or greater than the length of the head and thorax combined) and thread-like. **3**

3A   Hind legs flattened and fringed with long hairs. Body oval. .... **Dytiscidae**

3B   Hind legs not flattened and fringed with long hairs. Body usually parallel-sided, sometimes widening towards the apex. **4**

4A   Head at eyes as wide or wider than the pronotum. Elytra often patterned but without ridges and/or rows of punctations. .... **Cicindelidae**

4B   Head at eyes narrower than the pronotum. Elytra seldom patterned, but usually have ridges and/or rows of punctations. .... **Carabidae**

5A   Head extended forward into a snout, which may vary from short and blunt to long and slender. Antennae elbowed. .... **Curculionidae**

5B   Head not extended into a snout. Antennae variable. **6**

6A   Elytra short, exposing one or more segments of the tip of the abdomen when viewed dorsally. **7**

6B   Elytra long, completely covering the abdomen. **10**

7A   Antennae thread-like, not thickened at the tip. **8**

7B   Antennae clubbed, thickened or expanded in some manner at the tip. **9**

8A   Elytra approximately as long as pronotum. Head usually not wider than the pronotum. **Staphylinidae**

8B   Elytra much longer than pronotum. Head wider than pronotum. .... **Meloidae**

9A   Small (less than 10mm), with elytra squared off at the tips. .... **Histeridae**

9B   Small to medium-sized (5-35mm), but usually greater than 10mm. Tips of elytra are variable. .... **Silphidae**

10A Hind legs flattened and fringed with long hairs. Backward-pointing spine on venter between legs. .... **Hydrophilidae**

10B Hind legs not flattened and fringed with long hairs. No spine on venter between legs except in the family Elateridae. **11**

11A Tarsal formula appears 3-3-3. Body usually rounded or oval and extremely convex.
　　　　　　　　　　　　　　　　　　　　　　　　　　　　　Coccinellidae

11B Tarsal formula variable but not 3-3-3. Body variable. **12**

12A Tarsal formula 5-5-4. .... **13**

12B Tarsal formula appears 5-5-5 or 4-4-4. .... **14**

13A Pronotum narrower than both the head and the base of the elytra. Elytra soft and leathery, curving loosely around the body. .... **Meloidae**

13B Pronotum not narrower than both head and base of elytra. Elytra hard and shell-like.
　　　　　　　　　　　　　　　　　　　　　　　　　　　　　Tenebrionidae

14A Tarsal formula appears 4-4-4. .... **15**

14B Tarsal formula appears 5-5-5. .... **16**

15A Antennae long (equal to or greater than half the length of the body). Size variable but usually greater than 12mm. .... **Cerambycidae**

15B Antennae short (less than half the length of the body). Small (less than 12mm).
　　　　　　　　　　　　　　　　　　　　　　　　　　　　　Chrysomelidae

16A Elytra leathery, with reticulate network of veins between pronounced longitudinal ridges. Often brightly colored. .... **Lycidae**

16B Elytra variable, but without reticulate veins between longitudinal ridges. **17**

17A Posterior corners of pronotum pointed. Backward-pointing spine present between front two pairs of legs. .... **Elateridae**

17B Posterior corners of pronotum not pointed. No spine present between front two pairs of legs. **18**

18A Antennae thread-like or sawtoothed. **19**

18B Antennae clubbed or broadened at the tip. **21**

19A Body bullet-shaped, often with a metallic sheen. Elytra hard and shell-like. **Buprestidae**

19B Body elongate and parallel-sided, without a metallic sheen. Elytra leathery. **20**

20A   Head concealed by pronotum when viewed dorsally. Tip of abdomen yellow or white in color when viewed ventrally. .... **Lampyridae**

20B   Head typically not concealed by pronotum when viewed dorsally. Tip of abdomen same color as rest of body when viewed ventrally. .... **Cantharidae**

21A   Tip of antennae with expanded segments that form a symmetrical club. **22**

21B   Tip of antennae with extended projecting segments that form an asymmetrical club. **24**

22A   Elytra flattened and leathery, sometimes with ridges. .... **Silphidae**

22B   Elytra hard and shell-like. **23**

23A   Pronotum narrower than elytra at base. Body elongate and covered with hairs, often colorful. .... **Cleridae**

23B   Pronotum approximately the same width as elytra at base. Body rounded or oval, often with pigmented scales. .... **Dermestidae**

24A   Large (30-40mm), body wide and parallel-sided. Elytra with longitudinal grooves. **Passalidae**

24B   Size and body variable. Elytra without longitudinal grooves. **25**

25A   Antennae elbowed. Pronotum separated from base of elytra. .... **Lucanidae**

25B   Antennae not elbowed. Pronotum not separated from base of elytra. .... **Scarabaeidae**

The order Coleoptera contains more described insects than any other group. Beetles are incredibly diverse, and sometimes even bizarre. The harlequin beetle (*Acrocinus longimanus*) is unique in its vivid colors and alien appearing extended front legs. It is an excellent example of the insect wonders that can be found in the tropical rainforests of Central and South America.

Order Strepsiptera
Twisted-Winged Parasites

Characteristics of Strepsiptera

Small (4mm or less)
Sexes markedly different in appearance (dimorphic).
Development is via hypermetamorphosis.

Males

Forewings reduced to small knobs or clubs.
Hindwings large and fan-like.
Antennae with large lateral projections.
Eyes usually large and bulging.
Tarsi with 2-5 segments.
Chewing mouthparts.
Free-living and non-parasitic as adults.

Females

Wings absent.
Antennae absent.
Eyes absent.
Legs absent.
Mouthparts absent.
Live parasitically in the bodies of certain Homoptera and Hymenoptera.

**Fig. 462   Photomicrograph of an adult male twisted-winged parasite.
Order Strepsiptera**

**Fig. 463   Photomicrograph of an adult male twisted-winged parasite.
Order Strepsiptera**

Order Lepidoptera
Butterflies and Moths

Characteristics of Lepidoptera

Size extremely variable.
Two pairs of membranous wings covered with scales (Fig. 464).
Forewings slightly larger than the hindwings.
Antennae usually knobbed, thread-like, or feathery.
Labial palpi often large and obvious.
Mouthparts of the sucking type, consisting of a long slender tube
that remains coiled beneath the head when not in use (Fig. 465).
Complete metamorphosis.

Note: Many of the identifying characters of the Lepidoptera are related to their wing venation. Scales typically need to be removed from the wings to clearly observe the underlying venation. Since most specimens available to students have not been 'cleared' (Fig. 466), the treatment of Lepidoptera in this guide will not emphasize characters of wing venation.

It is convenient to group the Lepidoptera into the two broad, non-taxonomic categories of butterflies and moths, with which most of us are familiar. The families included here have been divided in that manner.

Terminology

frenulum - Spine or group of spines at the base of the hindwing that extends beneath the forewing and serves to hold both wings together.
jugum - Lobe at the base of the forewing that extends over the hindwing and serves to hold both wings together.
clearing - Process of removing the scales from lepidopteran wings by means of submersion in a bleach solution.

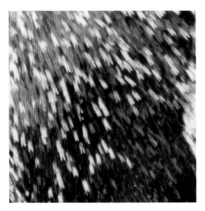

Fig. 464 Colors on the wing of a cecropia moth are caused by the combination of thousands of tiny scales.

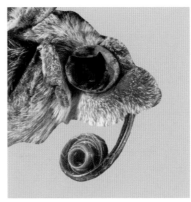

Fig. 465 The typical lepidopteran proboscis is seen in the curled tongue of a sphinx moth.

Fig. 466 A zebra butterfly illustrating the effect of chemically clearing the wings on one side to remove the scales and show the veins.

Order Lepidoptera
Butterflies and Moths

Butterflies

Characteristics of Butterflies
Antennae knobbed or clubbed.
Wings usually held vertically above the body when at rest.
Most species diurnal.
Most species pupate as a suspended chrysalis.

Fig. 467   A gulf fritillary showing the typical knobbed antennae found on butterflies.

Family Papilionidae - *Swallowtails*   (Figs. 468-470)
Hindwing of most species elongated into a tail-like extension.
Large (75-150mm).
Color variable.

Family Pieridae - *Sulfurs, Whites, & Orangetips*   (Figs. 471-473)
Medium-sized (20-75mm).
Usually yellow, orange, or white with some dark markings.
Tarsal claws forked.

Family Lycaenidae - *Hairstreaks, Coppers, & Blues*
Medium-sized (20-50mm).   (Figs. 474-476)
Antennae usually banded.
Hindwings typically with one or more thread-like extensions that may resemble antennae.
Front legs of males may be reduced.
Many species with bright or iridescent colors.

Family Danaidae - *Milkweed Butterflies*   (Figs. 477-479)
Large (75-100mm).
Front legs greatly reduced.
Wings usually orange or brownish, with black and white markings.

Fig. 468   Zebra swallowtail butterfly.
Family Papilionidae

Fig. 469   Giant swallowtail butterfly.
Family Papilionidae

Fig. 470   Tiger swallowtail butterfly.
Family Papilionidae

Fig. 471   Cabbage butterfly.
Family Pieridae

Fig. 472   Alfalfa butterfly.
Family Pieridae

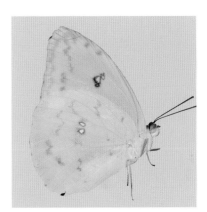

Fig. 473   Cloudless sulfur butterfly.
Family Pieridae

Fig. 474   Hairstreak butterfly upper
surface. Note the tails and banded
antennae.   Family Lycaenidae

Fig. 475   Hairstreak butterfly under
surface. Note the tails and spots.
Family Lycaenidae

Fig. 476   The Florida atala butterfly, a
large tailless lycaenid butterfly.
Family Lycaenidae

Fig. 477   Queen butterfly showing
typical danaid markings.
Family Danaidae

Fig. 478   A monarch butterfly, one of
the most common U.S. danaids.
Family Danaidae

Fig. 479   Tropical danaid still exhibits the
characteristic wing shape and markings.
Family Danaidae

Order Lepidoptera
Butterflies and Moths

Butterflies

Family Nymphalidae - *Brush-Footed Butterflies*   (Figs. 480-482)
Front legs greatly reduced.
Veins in the forewings are not greatly enlarged at their base.
Size and color variable.

Family Satyridae - *Satyrs & Nymphs*   (Figs. 483-485)
Small to medium-sized (20-40mm).
Front legs greatly reduced.
Some veins in the forewings greatly enlarged or inflated at their base.
Color often a dull brown or gray, commonly pigmented with eyespots.

Family Heliconidae - *Heliconians*   (Figs. 486-488)
Medium-large (60-100mm).
Forewings long and narrow, almost oval.
Eyes large.
Antennae long.
Typically brightly pigmented.

Skippers

Note: The skippers are considered to be in their own group, apart from the butterflies.

Family Hesperiidae - *Skippers*   (Figs. 489-491)
Antennae hooked or curved at the tips.
Body thick and stout.
Some species with tail-like extensions on the hindwings.

Fig. 480 The viceroy butterfly has colors and markings similar to the monarch. **Family Nymphalidae**

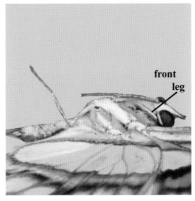

Fig. 481 The front legs of brush-footed butterflies are greatly reduced. **Family Nymphalidae**

Fig. 482 Red admiral butterfly. **Family Nymphalidae**

Fig. 483 Satyrid butterflies are often dull-colored with eyespots like this wood nymph. **Family Satyridae**

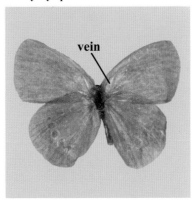

Fig. 484 The veins of the forewings of satyrid butterflies may be greatly enlarged at their base. **Family Satyridae**

Fig. 485 The forewing above shows the typical expanded base of the wing veins. **Family Satyridae**

Fig. 486 A julia butterfly shows the typical long oval wings found on heliconians. **Family Heliconidae**

Fig. 487 Tropical heliconian butterfly. **Family Heliconidae**

Fig. 488 The zebra longwing butterfly exhibits the typical heliconian features. **Family Heliconidae**

Fig. 489 Longtailed skipper butterfly. **Family Hesperiidae**

Fig. 490 The antennae of skipper butterflies are hooked at the tips. **Family Hesperiidae**

Fig. 491 A Brazilian skipper shows the stout body that is typical of the skippers. **Family Hesperiidae**

Order Lepidoptera
Butterflies and Moths

Moths

Fig. 492 A cecropia moth showing the typical feathery antenna and robust body of many moths.
Family Saturniidae

Characteristics of Moths

Antennae feathery or thread-like.
Wings usually held horizontally out from the body when at rest (sometimes with hindwings concealed).
Most species nocturnal.
Most species pupate either within a silken cocoon or as a naked pupa in a protected place.

Family Sphingidae - *Sphinx or Hawk Moths*   (Figs. 493-495)

Medium-large (25-125mm).
Body large, thick, and torpedo-shaped.
Forewings pointed and narrowed, about twice the size of the hindwings.
Antennae usually thread-like, but thickened apically.

Family Saturniidae - *Giant Silkworm Moths*   (Figs. 496-501)

Medium-large (30-175mm).
Wings usually contain an eyespot or clear, window-like area.
Antennae feathery.
Body usually hairy, and does not extend past the hindwings.

Family Citheroniidae - *Royal Moths*   (Figs. 502-503)

Medium-large (30-175mm).
Wings usually without eyespots or large clear areas.
Antennae feathery only at basal portion.
Body hairy and extends beyond the hindwings.
Tongue reduced or completely absent.

Family Pterophoridae - *Plume Moths*   (Fig. 504)

Small-medium (14-40mm).
Wings divided into lobes (forewings with 2-4, hindwings with 3).
T-shaped appearance when viewed dorsally.
Legs long (approximately the length of the body).
Light-colored brown, gray, or white.

Fig. 493   Carolina sphinx moth showing the characteristic wing and body shape of hawk moths.   Family Sphingidae

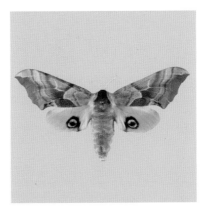

Fig. 494   One-eyed sphinx moth. Note the large, torpedo-shaped body. Family Sphingidae

Fig. 495   The hummingbird sphinx moth has wings that are nearly devoid of scales. Family Sphingidae

Fig. 496   The extremely feathery antennae of a male polyphemus moth. Family Saturniidae

Fig. 497   The antennae of female giant silkworm moths, like this polyphemus, are usually less feathery.   Family Saturniidae

Fig. 498   Female io moth. The fore-wings normally conceal the eyespots on the hindwings.   Family Saturniidae

Fig. 499   The luna moth has transparent 'windows' in the wings that are typical of giant silk moths.   Family Saturniidae

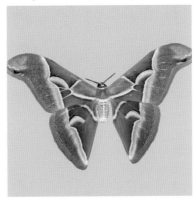

Fig. 500   The ailanthus silk moth or cynthia moth was introduced into the U.S. from China.   Family Saturniidae

Fig. 501   The cecropia or robin moth. Family Saturniidae

Fig. 502   The royal walnut moth is the adult stage of the hickory horned devil caterpillar.   Family Citheroniidae

Fig. 503   The imperial moth, a large common royal moth found in the U.S. Family Citheroniidae

Fig. 504   The wings of plume moths are divided into distinct lobes. Family Pterophoridae

135

Order Lepidoptera
Butterflies and Moths

Moths

Family Sesiidae - *Clear-Winged Moths*   (Fig. 505)
Medium-sized (15-45mm).
Large areas of the wings transparent and without scales.
Forewings long and narrow.
Hind edge of forewing and front edge of hindwing with a row of very small spines at the base.
Many species extremely wasp-like in appearance.

Family Arctiidae - *Tiger Moths*   (Figs. 506-507)
Medium-sized (15-70mm).
Body usually hairy.
Antennae slightly feathery or thickened.
Most species with a light base color with bright or dark markings on the wings and/or abdomen.

Family Ctenuchidae - *Wasp Moths*   (Figs. 508-510)
Medium-sized (25-50mm).
Ocelli present.
Often dark or metallic in color with yellow or red on the wings.
Very wasp-like in appearance and behavior.

Family Noctuidae - *Noctuid Moths*   (Figs. 511-513)
Size variable (15-170mm).
Antennae thread-like.
Ocelli usually present.
Forewings usually dark with a mottled pattern of lines or dots.
Hindwings may match forewings in color or have bright colors that are concealed when at rest.

Family Geometridae - *Geometer Moths*   (Figs. 514-516)
Small-medium (10-60mm).
Antennae variable.
Ocelli usually absent.
Forewings and hindwings usually similar in color and pattern.

Fig. 505 Clear-winged moths may be wasp-like with large areas of the wings devoid of scales.   Family Sesiidae

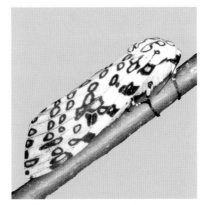

Fig. 506 The giant leopard moth shows the typical light base coloring with dark markings.   Family Arctiidae

Fig. 507 The virgin tiger moth. Family Arctiidae

Fig. 508 Scarlet-bodied wasp moth. Family Ctenuchidae

Fig. 509 Patriot or polka-dot wasp moth.
Family Ctenuchidae

Fig. 510 The moth of the spotted oleander caterpillar.
Family Ctenuchidae

Fig. 511 The underwing moth is a noctuid with bright colors on the hindwings.   Family Noctuidae

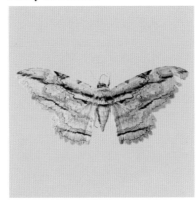

Fig. 512 The owl moth is a cryptic noctuid where the forewings and hind-wings match.   Family Noctuidae

Fig. 513 The black witch is one of the largest domestic noctuid moths. Family Noctuidae

Fig. 514 Southern pine looper moth. Family Geometridae

Fig. 515 Oak besma, a geometer moth. Family Geometridae

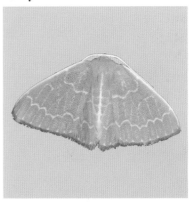

Fig. 516 Geometer moths are often cryptic and well camouflaged. Family Geometridae

Key to the Families of Lepidoptera

1A   Antennae hooked at the tip. Body thick and stout. .... **Hesperiidae**

1B   Antennae not hooked at the tip. Body variable. **2**

2A   Antennae with a knob at the tip. (**Butterflies**) **3**

2B   Antennae feathery, thread-like, or thickened. (**Moths**) **9**

3A   Front legs the same size as other legs, or only slightly reduced. **4**

3B   Front legs greatly reduced. **6**

4A   Hindwings with tail-like or thread-like extensions. Tarsal claws not forked. **5**

4B   Hindwings without tail-like or thread-like extensions. Tarsal claws forked. .... **Pieridae**

5A   Large (75-100mm). Antennae not banded. .... **Papilionidae**

5B   Medium-sized (15-50mm). Antennae often banded. .... **Lycaenidae**

6A   Large (75-100mm). Wings usually orange or brownish with black and white markings
along the edges. No scales on antennae. .... **Danaidae**

6B   Size variable. Wings variable, but not as described above. Antennae with scales. **7**

7A   Small to medium-sized (20-40mm). Wings usually dull gray or brown with small eyespots.
Some of the veins of the forewing are greatly enlarged at their base. .... **Satyridae**

7B   Size variable. Color variable, but not as described above. No veins of forewing greatly
enlarged at their base. **8**

8A   Large (60-100mm). Forewings long and oval, their span almost twice as long as the
hindwings. .... **Heliconidae**

8B   Size variable. Forewings not long and oval, approximately the same length as the
hindwings. .... **Nymphalidae**

9A   Small to medium-sized (15-40mm). Wings divided into narrow lobes (forewings with 2-4,
hindwings with 3). .... **Pterophoridae**

9B   Size variable. Wings not divided into narrow lobes. **10**

10A   Medium-sized to large (25-125mm). Forewing pointed and almost triangular in shape, approximately twice as long and twice as big as hindwing. Body thick and robust, with antennae thickened apically. .... **Sphingidae**

10B   Size variable. Forewings variable, but not as described above. Body and antennae variable, but not as described above. **11**

11A   Extremely wasp-like in appearance. Medium-sized (15-50mm). **12**

11B   Appearance variable, but not wasp-like. Size variable. **13**

12A   Transparent areas on all wings, especially the hindwings. Forewings narrow throughout their length and almost parallel-sided. .... **Sesiidae**

12B   Wings variable, but mostly without transparent areas. Usually pigmented with bright or metallic colors. Forewings not parallel-sided, but narrow at base. .... **Ctenuchidae**

13A   Medium-sized to large, with thick and extremely hairy bodies. Wings colorful, often with spots, bars, eyespots, or 'windows' (small transparent areas). **14**

13B   Size variable, with tapering bodies that are not extremely hairy. Wings with subdued or cryptic colors. **16**

14A   Ocelli present. Body usually with rows of spots. .... **Arctiidae**

14B   Ocelli absent. Body variable. **15**

15A   Antennae feathery. Wings with eyespots or windows. Abdomen shorter than hindwings (when wings are spread). .... **Saturniidae**

15B   Antennae variable. Wings without eyespots or windows. Abdomen longer than hindwings (when wings spread). .... **Citheroniidae**

16A   Ocelli usually present. Labial palps long. Antennae thread-like. .... **Noctuidae**

16B   Ocelli usually absent. Labial palps short. Antennae variable. .... **Geometridae**

Order Trichoptera
Caddisflies

Characteristics of Trichoptera

Small-medium (5-25mm).

Four membranous wings usually present.

Wings with only a few cross veins.

Wings all approximately the same size and held roof-like over the body.

Body and wings hairy.

Antennae thread-like and long (equal to or greater than the length of the body).

Tarsi with five segments.

Resemble moths in appearance.

Larvae aquatic and construct protective cases of various materials.

Sponging mouthparts, with long palpi.

Complete metamorphosis.

Fig. 517   Lateral view of an adult caddisfly.
Order Trichoptera

Fig. 518   Oblique view of an adult caddisfly.
Family Hydropsychidae
Courtesy BP Stark

Order Diptera
Flies

Characteristics of Diptera

One pair of wings (forewings), with halteres in place of hindwings.
Antennae variable.
Sucking, sponging, and/or piercing mouthparts.
Complete metamorphosis.

Terminology

haltere - Small knobbed structure on each side of the metathorax, replacing the hindwings.
calypters - Two lobes at the base of each wing near the anal margin.
arista - Large bristle or hair protruding from the third (last) antennal segment of certain flies.
pulvilli - Pads or lobes located beneath the tarsal claws.
scutellum - Sclerite that is roughly triangular in shape located at the posterior of the thorax.
postscutellum - Area just behind or below the scutellum.
hypopleuron - Sclerite on the thorax just above the hind coxa.
frontal suture - Suture in the shape of an inverted 'U' on the face of some flies.

**Fig. 519   The horsefly is a typical representative of the Order Diptera.
Family Tabanidae**

**Fig. 520   The papaya fruit fly has an unusually long ovipositor.
Family Tephritidae**

Suborder Nematocera
Nematocerous Flies

Characteristics of Nematocera

Antennae with six segments or more.
Antennae long (equal to or greater than the length of the the head and pronotum combined).
Often mosquito-like in appearance.

Family Culicidae - *Mosquitoes*   (Figs. 521-523)

Scales on the veins and margins of the wings, and sometimes on the body.
Wings long and narrow.
Mouthparts consist of a long, slender beak or proboscis.
Wingtip with a straight, unbranched vein reaching the margin between two branched veins.

Family Tipulidae - *Crane Flies*   (Figs. 524-526)

Medium-sized (20mm or more).
Resemble giant mosquitoes.
Wings long and narrow, often pigmented but without scales.
Legs extremely long (almost twice the length of the body).
Top of thorax with a V-shaped suture present.

Family Chironomidae - *Midges*   (Fig. 527)

Resemble mosquitoes in general appearance.
Wings long and narrow, but without scales.
Wingtip without a straight vein reaching margin between two branched veins.
Front tarsi may be extremely long.

Family Psychodidae - *Moth & Sand Flies*   (Fig. 528-529)

Small (5mm or less).
Body and wings densely covered with hairs.
Wings usually pointed at the tips.
Resemble tiny moths.

Family Simuliidae - *Black Flies or Buffalo Gnats*   (Fig. 530)

Small (5mm or less).
Body is humpbacked when viewed in profile.
Anterior wing veins near the costal margin are heavier and thicker than the other wing veins.
Body color is usually black or gray.

Family Bibionidae - *March Flies & Lovebugs*   (Figs. 531-532)

Body longer than 5mm.
Antennae short (less than the length of the head and the thorax combined).
Tibiae with apical spurs present.
Costal margin of the wing remains heavy and thickened almost to the wingtip.

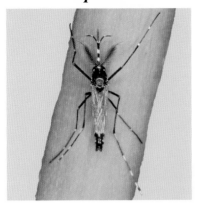

Fig. 521   Male mosquitoes have plumose antennae.   Family Culicidae
Photo JL Castner   Courtesy IFAS/UF

Fig. 522   Mosquito wingtip showing the scales that line the veins and wing edges. Family Culicidae

Fig. 523   Adult mosquito in the genus *Psorophora*. Family Culicidae

Fig. 524   A crane fly resembles a large mosquito and often has pigmented wings. Family Tipulidae

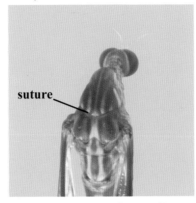

Fig. 525   Dorsal view of a crane fly thorax showing the V-shaped suture. Family Tipulidae

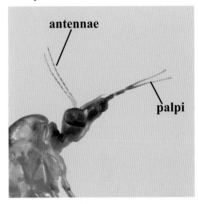

Fig. 526   Head of a crane fly showing the many-segmented antennae and long palpi.   Family Tipulidae

Fig. 527   Midges or 'blind mosquitoes' have clear wings without scales. Family Chironomidae

Fig. 528   Moth flies have extremely hairy wings and bodies. Family Psychodidae

Fig. 529   Lateral view of a moth fly showing the distinctive wing and body shape.   Family Psychodidae

Fig. 530   Black fly or buffalo gnat. Family Simuliidae

Fig. 531   Adult female lovebug. Family Bibionidae

Fig. 532   Head and antennae of a lovebug, a species of march fly. Family Bibionidae

Suborder Brachycera
Brachycerous Flies

Characteristics of Brachycera

Medium-large flies.

Antennae short (seldom as long as the head and thorax combined), with 3-5 segments.

Antennae seldom aristate, but may have a long slender terminal style extending from the last segment.

Stout-bodied flies that do not resemble mosquitoes.

No frontal suture present.

Family Asilidae - *Robber Flies*   (Figs. 533-535)

Hair on head gives face a 'bearded' appearance.

Top of head with a depression between the eyes, three ocelli present.

Antennae with three segments and a terminal style.

Body with elongate abdomen.

Pulvilli often present and obvious.

Some species resemble bees.

Family Mydidae - *Mydas Flies*   (Figs. 536-537)

Medium-large (25mm or more).

Antennae with four segments, the last of which is enlarged.

Either one ocellus or none.

Wingtip with three nearly parallel cells of the same approximate size.

Tip of mouthparts has two fleshy lobes.

Family Stratiomyidae - *Soldier Flies*   (Fig. 538)

Antennae with three segments, the last of which is rounded or elongate.

Abdomen usually wide and flattened.

Some species resemble wasps.

Family Tabanidae - *Horse & Deer Flies*   (Figs. 539-541)

Medium-large (15-25mm).

Third (last) antennal segment elongate and ringed, sometimes with a basal tooth.

Wingtip enclosed by a forked vein.

Calypters of wings large.

Smaller species often with banded or pigmented wings.

Family Bombyliidae - *Bee Flies*   (Figs. 542-544)

Stout-bodied flies, often hairy.

Third (last) antennal segment variable.

Two wavy, nearly parallel veins near wingtip.

Some species with a long slender proboscis and/or patterned wings.

Fig. 533 Bee-mimicking robber fly. Note the bearded appearance of the face. Family Asilidae

Fig. 534 Lateral view of a robber fly, showing the characteristic tapering abdomen. Family Asilidae

Fig. 535 Typical robber fly. Note the beak and the pulvilli visible beneath the tarsi. Family Asilidae

Fig. 536 Dorsal view of a large, wasp-like mydas fly. Family Mydidae

Fig. 537 Wasp-mimicking mydas fly. Note the characteristic antennae. Family Mydidae

Fig. 538 Wasp-mimicking soldier fly. Family Stratiomyidae

Fig. 539 Deer fly. Family Tabanidae

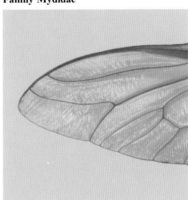

Fig. 540 Wing of a horse fly showing how the tip is enclosed by a forked vein. Family Tabanidae

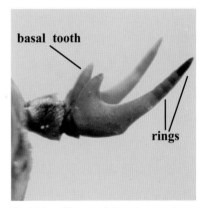

basal tooth

rings

Fig. 541 Characteristic antennae of a horse fly. Family Tabanidae
Photo JL Castner   Courtesy IFAS/UF

Fig. 542 Dorsal view of a bee fly. Note the patterned wings. Family Bombyliidae

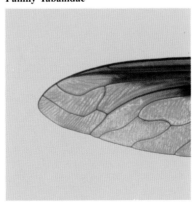

Fig. 543 Venation in the wingtip of a bee fly. Family Bombyliidae

Fig. 544 Extremely furry bee fly with a long proboscis. Family Bombyliidae

145

Suborder Cyclorrhapha
Cyclorrhaphous Flies

Characteristics of Cyclorrhapha

Antennae aristate (three segments with a hair or arista on the third).
Stout-bodied and do not resemble mosquitoes.
Frontal suture present or absent.

Family Syrphidae - *Hover Flies*   (Figs. 545-547)

Spurious vein present near middle of wing.
Antennae with dorsal arista.
Long narrow, often irregularly-shaped cell terminating near or extending to wingtip.
Proboscis thick and fleshy.
Face usually convex, frontal suture absent.
Some species resemble bees and wasps, frequently yellow and black.

Family Conopidae - *Thick-Headed Flies*   (Figs. 548-549)

Head wider than thorax.
Antennae approximately as long as the head and thorax combined.
No spurious vein.
Proboscis slender, often projects straight out in front of the head.
Abdomen narrower at the base than the tip.
Frontal suture absent.
Some species resemble thread-waisted wasps.

Family Phoridae - *Scuttle Flies*   (Fig. 550)

Small (6mm or less) and humpbacked when viewed in profile.
Veins strong and heavy near the basal half of the anterior edge of the wing.
Hind femora flattened.

Family Tephritidae - *Fruit Flies*   (Figs. 551-553)

Small-medium flies.
Wings often banded or pigmented.
First vein below the anterior edge of the wing near the base, bends at a right angle to the edge,
but doesn't reach it.
Closed anal cell present that has a posteriorly pointing portion.

Family Drosophilidae - *Vinegar Flies*   (Figs. 554-556)

Small (4mm or less).
Anterior edge of wing is thickened and broken at two points near where wing veins join it.
Arista of antennae plumose.
Face appears to be ridged.

Fig. 545   Wasp-mimicking hover fly.
Family Syrphidae

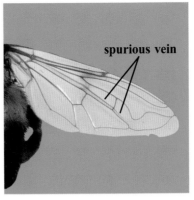

Fig. 546   Wing venation of a hover fly showing the spurious vein.
Family Syrphidae

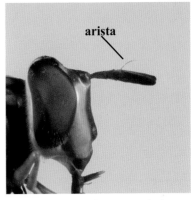

Fig. 547   Head and antennae of a hover fly.
Family Syrphidae

Fig. 548   Dorsal view of a wasp-like thick-headed fly.
Family Conopidae

Fig. 549   Lateral view of a thick-headed fly showing the typical shape.
Family Conopidae

Fig. 550   Humpbacked or scuttle fly.
Family Phoridae

Fig. 551   Goldenrod gall fly.
Family Tephritidae

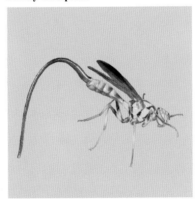

Fig. 552   Female papaya fruit fly.
Family Tephritidae

Fig. 553   Caribbean fruit fly.
Family Tephritidae

Fig. 554   Vinegar fly.
Family Drosophilidae

Fig. 555   Wing venation of a vinegar fly.
Family Drosophilidae

Fig. 556   Vinegar fly on a pine needle.
Family Drosophilidae

Suborder Cyclorrhapha
Cyclorrhaphous Flies

Family Tachinidae - *Tachinid Flies*   (Figs. 557-559)

Arista completely bare.
Usually stout-bodied, bristly flies.
Large cell nearest wingtip narrows distally.
Postscutellum visible as a prominent lobe beneath the scutellum.
Dorsal portion of abdomen often overlaps ventral portion.
Hypopleura with vertical series of bristles.
Pteropleura with bristles.
Frontal suture present.

Family Calliphoridae - *Blow Flies & Bottle Flies*

Arista plumose throughout its length.
Body often metallic (sometimes only showing a dull metallic sheen).   (Figs. 560-562)
Large cell nearest wingtip narrows distally.
Postscutellum not well developed.
Hypopleura with bristles.
Pteropleura with bristles.
Two notopleural bristles present.
Frontal suture present.

Family Sarcophagidae - *Flesh Flies*   (Figs. 563-565)

Arista plumose on its basal portion and bare on its distal portion.
Thorax with a pattern of gray and black longitudinal stripes.
Abdomen with a gray and black checkered pattern.
Large cell nearest wingtip narrows distally.
Four notopleural bristles present.
Frontal suture present.

Family Muscidae - *House Flies & Others*   (Figs. 566-568)

Arista plumose throughout its length.
Thorax often striped while the sides of the abdomen are pale.
Large cell nearest wingtip may narrow distally or remain parallel-sided.
Hypopleura bare or with weak hairs rather than strong bristles.
Pteropleura bare or with weak hairs rather than strong bristles.
Frontal suture present.

Fig. 557   Tachinid flies are often stout-bodied and hairy.
Family Tachinidae

Fig. 558   The arista on the antenna of a tachinid fly is completely bare.
Family Tachinidae

Fig. 559   Feather-legged fly, a parasite of the green stink bug.
Family Tachinidae

Fig. 560   A greenbottle fly showing the characteristic metallic sheen.
Family Calliphoridae

Fig. 561   The arista on the antenna of a blow fly is plumose throughout.
Family Calliphoridae

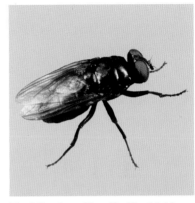

Fig. 562   Some blow flies like this blue-bottle are very shiny while others have a dull metallic glint.   Family Calliphoridae

Fig. 563   Flesh fly showing the striped thorax and checkered abdomen.
Family Sarcophagidae

Fig. 564   The arista on the antenna of a flesh fly is plumose at the base and bare at the tip.   Family Sarcophagidae

Fig. 565   Red-tailed flesh fly.
Family Sarcophagidae

Fig. 566   The house fly is one of the most common muscid flies.
Family Muscidae

Fig. 567   The arista on the antenna of a muscid fly is plumose throughout.
Family Muscidae

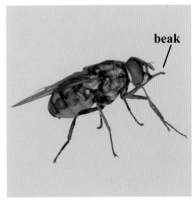

Fig. 568   The stable fly is a biting muscid fly. Note the prominent beak.
Family Muscidae

Key to the Families of Flies

1A   Antennae with six or more segments. (**Suborder Nematocera**) **2**

1B   Antennae with five or less segments. (Note: The third antennal segment of some groups may be ringed, giving the impression of multiple segments.) **7**

2A   Wings with scales or hairs on the veins and edges. **3**

2B   Wings without scales or hairs on the veins and edges. **4**

3A   Proboscis long (equal to or greater than the length of the antennae). Wingtip rounded. Size variable. .... **Culicidae**

3B   Proboscis short (much less than the length of the antennae). Wingtip pointed. Small (5mm or less). .... **Psychodidae**

4A   Body and wings mosquito-like in appearance. No long proboscis present. **5**

4B   Body and wings not mosquito-like. No long proboscis present. **6**

5A   Medium-sized (15-25mm). V-shaped suture on dorsum of thorax. All legs very long.   **Tipulidae**

5B   Small (2-10mm). No V-shaped suture on dorsum of thorax. Front legs may be long, but never all legs very long. .... **Chironomidae**

6A   Ocelli absent. Humpbacked in appearance when viewed laterally. Small (4mm or less).   **Simuliidae**

6B   Ocelli present. Not particularly humpbacked in appearance when viewed laterally. Small to medium-sized (10-15mm). .... **Bibionidae**

7A   Antennae with four segments. Last segment enlarged and thickened. Large (25mm or more). .... **Mydidae**

7B   Antennae with three segments. Size variable. **8**

8A   Antennae without arista, or if arista present then third antennal segment with rings. (**Suborder Brachycera**) **9**

8B   Antennae with arista. Third antennal segment without rings. (**Suborder Cyclorrhapha**) **12**

9A   Face extremely hairy and bearded in appearance. Top of head with depression between the eyes. .... **Asilidae**

9B   Face not hairy or bearded in appearance. Top of head without a depression between the eyes. .... **Stratiomyidae**

10A   Wingtip enclosed by a forked vein. Antennae may have basal tooth. .... **Tabanidae**
10B   Wingtip not enclosed by a forked vein. Antennae never with a basal tooth. **11**

11A   Wing with two wavy, nearly parallel veins that terminate before the tip.
          Usually stout-bodied and hairy. Some species with a very long proboscis.   **Bombyliidae**
11B   Wing without two wavy, nearly parallel veins near tip. Usually slender-bodied and
          sometimes wasp-like. Never with a very long proboscis. .... **Stratiomyidae**

12A   Small (5mm or less). **13**
12B   Size variable, but greater than 5mm. **14**

13A   Humpbacked in appearance when viewed laterally. Veins heavy around proximal portion
          of costal edge of wing, while others reduced and directed obliquely across wing.
          Arista bare. .... **Phoridae**
13B   Not humpbacked in appearance when viewed laterally. Costal vein broken in two places
          near base, while others not reduced and are directed straight through wing parallel to the
          costa. Arista plumose. .... **Drosophilidae**

14A   Wings banded or spotted. Subcosta vein bends up towards costa at almost a right angle,
          but doesn't reach it. .... **Tephritidae**
14B   Wings variable. Subcosta does not bend up at right angles towards costa. **15**

15A   Frontal suture absent. **16**
15B   Frontal suture present. **17**

16A   Proboscis long and slender. Tip of abdomen thickened and base of abdomen narrowed.
          No spurious vein present. .... **Conopidae**
16B   Proboscis short and fleshy. Abdomen variable. Spurious vein present near middle
          of wing. .... **Syrphidae**

17A   Arista completely bare. Postscutellum visible as a prominent lobe beneath the scutellum.
                                                                                                               Tachinidae
17B   Arista partially or entirely plumose. Postscutellum not well developed. **18**

18A   Only basal half of arista plumose. Body not metallic. Thorax with black and gray stripes.
                                                                                                               Sarcophagidae
18B   Arista entirely plumose. Body and thorax variable. **19**

19A   Body metallic. Hypopleura and pteropleura with strong bristles. .... **Calliphoridae**
19B   Body not metallic. Hypopleura and pteropleura bare or with weak hairs. .... **Muscidae**

Order Siphonaptera
Fleas

Characteristics of Siphonaptera

Small (8mm or less).
Wings absent.
Extremely compressed laterally.
Antennae short (less than the length of the head)
and usually concealed.
Compound eyes present or absent.
Ocelli absent.
Coxae very large and long.
Tarsi with five segments.
Piercing-sucking mouthparts.
Complete metamorphosis.

Terminology

genal comb - Row of large dark spines at the bottom of the front of the head
pronotal comb - Row of large dark spines along the posterior edge of the pronotum.

Family Pulicidae - *Common Fleas*

Genal comb present or absent.
Pronotal comb present or absent.
Compound eyes present and well developed.
Abdominal terga 2-6 with a single transverse row of bristles.

Family Dolichopsyllidae - *Rodent Fleas*

Genal comb usually absent.
Pronotal comb present.
Compound eyes absent or greatly reduced.
Some of abdominal terga 2-6 with two transverse rows of bristles.

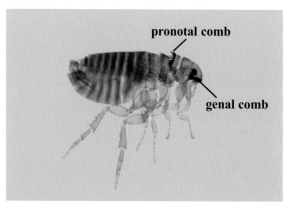

Fig. 569   Photomicrograph of a cat flea.
Family Pulicidae

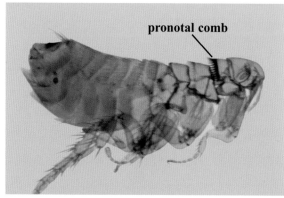

Fig. 570   Photomicrograph of a rodent flea.
Family Dolichopsyllidae

Order Hymenoptera
Ants, Bees, Wasps, and Others

Characteristics of Hymenoptera

Two pairs of wings or wingless.
Forewings larger than hindwings.
Antennae with 10 or more segments.
Antennae longer than head, but seldom longer than head and thorax combined.
Females with a well developed ovipositor (referred to as a 'stinger' if used for defense).
Tarsi usually with five segments.
Chewing or chewing-lapping mouthparts.
Complete metamorphosis.

Terminology

tegula - Plate-like sclerite that covers the base of the forewing.
stigma - Thickened dark spot along the front edge of the wing near the tip.
node - Dorsal hump found on the first and second abdominal segments of ants.
propodeum - First abdominal segment of Apocrita, fused to the thorax.
petiole - Slender stalk-like portion of the propodeum.
ovipositor - Egg-laying device, variable in form and size, located at the tip of the abdomen.
basal cells - Elongate closed cells originating at the base of the wing.
marginal cells - Cells that border the front edge of the forewing from a point beneath the stigma to the wingtip.

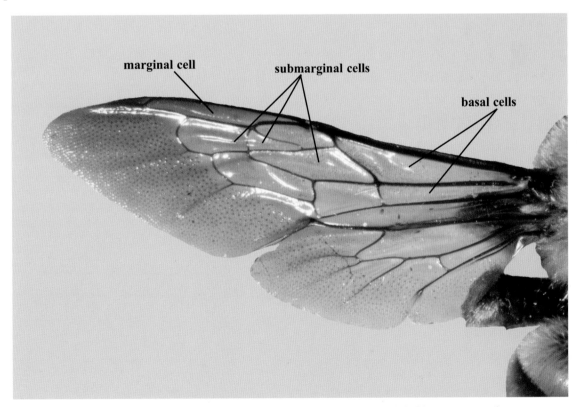

Fig. 571   Wing venation of a carpenter bee illustrating the various cells found in the typical hymenopteran wing.

153

Suborder Symphyta
Horntails and Sawflies

Characteristics of Symphyta

Abdomen broadly joined to the thorax. (Note: This character is difficult to see unless the wings of the specimen are spread.)
Hindwings with three basal cells.
Trochanters with two segments.

Family Siricidae - *Horntails*   (Figs. 572-574)

Large (25-35mm).
Dorsal spur at the tip of the abdomen.
Females with stout straight ovipositor projecting from abdomen beneath the dorsal spur.
Pronotum wider than long, narrowed at the center when viewed dorsally.
Front tibiae have one apical spur.

Family Tenthredinidae - *Common Sawflies*   (Figs. 575-577)

Size variable (20mm or less).
Antennae thread-like and usually with 9 segments.
Forewings with one or two marginal cells.
Front tibiae with two apical spurs.

Family Diprionidae - *Conifer Sawflies*   (Figs. 578-580)

Small (12mm or less).
Antennae sawtoothed or comb-like, with 13 segments or more.
Body wide when viewed dorsally.
Forewings with only one marginal cell.
Front tibiae with two apical spurs.

Family Cimbicidae - *Cimbicid Sawflies*   (Figs. 581-582)

Medium-sized (15-25mm).
Antennae clubbed, with 7 segments or less.
Front tibiae with two apical spurs.

Family Cephidae - *Stem Sawflies*   (Fig. 583)

Small-medium (20mm or less).
Antennae clubbed, with 17-30 segments.
Body laterally compressed.
Front tibiae with one or two apical spurs.

Fig. 572   Dorsal view of a male horntail.
Family Siricidae

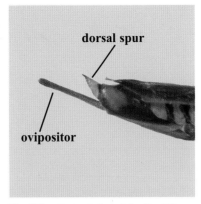

Fig. 573   Lateral view of the tip of the abdomen of a female horntail.
Family Siricidae

Fig. 574   Lateral view of the tip of the abdomen of a male horntail.
Family Siricidae

Fig. 575   Dorsal view of a common sawfly.
Family Tenthredinidae

Fig. 576   Dorsal view of a common sawfly.
Family Tenthredinidae

Fig. 577   Head and antennae of a common sawfly.
Family Tenthredinidae

Fig. 578   Dorsal view of a conifer sawfly.
Family Diprionidae

Fig. 579   Head and antennae of a male conifer sawfly.
Family Diprionidae

Fig. 580   Head and antennae of a female conifer sawfly.
Family Diprionidae

Fig. 581   Dorsal view of a cimbicid sawfly.
Family Cimbicidae

Fig. 582   Head and antennae of a cimbicid sawfly.
Family Cimbicidae

Fig. 583   Dorsal view of a stem sawfly.
Family Cephidae

Suborder Apocrita
Ants, Bees, and Wasps

Characteristics of Apocrita

Base of abdomen constricted or distinctly stalked.
Hindwing with two basal cells or less.
Trochanters with one or two segments.
Females with either an internal 'stinger' or a well developed external ovipositor.

Family Ichneumonidae - *Ichneumon Wasps*   (Figs. 584-586)

Color and size variable (5-40mm).
Body often laterally compressed.
Antennae thread-like, with 16 or more segments.
Antennae long (greater than or equal to half the body length).
Hind trochanters with two segments.
Forewings with a large, somewhat trapezoidal-shaped closed cell directly below the stigma.
Forewings sometimes with a small closed, almost round cell (areolet) beyond the stigma.
Ovipositor may be extremely long and thread-like.

Family Braconidae - *Braconid Wasps*   (Figs. 587-589)

Small-medium (5-15mm), often dark in color.
Antennae thread-like, with 16 or more segments.
Antennae long (greater than or equal to half the body length).
Hind trochanters with two segments.
Forewings with two small closed cells below the stigma.
Females of some species with long ovipositors.

Family Chalcididae - *Chalcidid Wasps*   (Figs. 590-592)

Small (10mm or less).
Hind femora greatly enlarged and serrate beneath.
Hind coxae larger and longer than front coxae.
Antennae elbowed, with 13 or less segments.
Venation of wings greatly reduced, except near the front margin.

Family Evaniidae - *Ensign Wasps*   (Figs. 593-595)

Small (10mm or less).
Thorax large and abdomen very small.
Abdomen attached to the thorax by a slender petiole that originates high above the hind coxae.
Legs long and spider-like.
Forewings with strong or reduced venation.

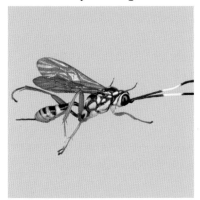

Fig. 584   Female ichneumon wasp. Family Ichneumonidae

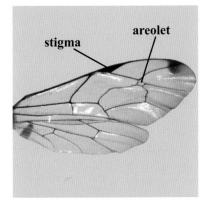

Fig. 585   Venation of an ichneumon wasp with an areolet. Family Ichneumonidae

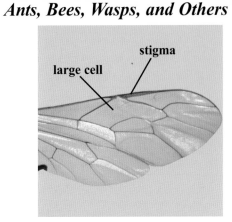

Fig. 586   Venation of an ichneumon wasp. Note the large cell beneath the stigma.   Family Ichneumonidae

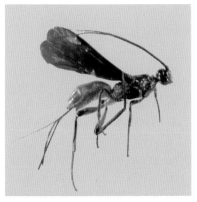

Fig. 587   Lateral view of a braconid wasp. Family Braconidae

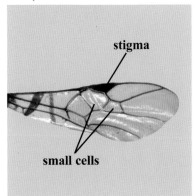

Fig. 588   Venation of a braconid wasp. Note the two small cells beneath the stigma. Family Braconidae

Fig. 589   Venation of a braconid wasp. Family Braconidae

Fig. 590   Lateral view of a chalcidid wasp. Note the enlarged hind femur. Family Chalcididae

Fig. 591   Chalcidid wasp. Family Chalcididae
Photo JL Castner   Courtesy IFAS/UF

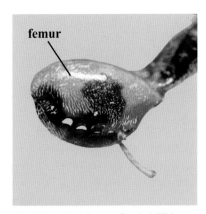

Fig. 592   Hind femur of a chalcidid wasp. Family Chalcididae

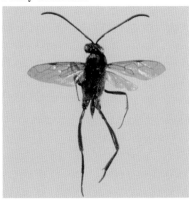

Fig. 593   Dorsal view of an ensign wasp. Note the long spidery legs. Family Evaniidae

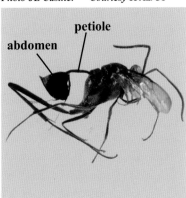

Fig. 594   Lateral view of a female ensign wasp. Note the small triangular abdomen.   Family Evaniidae

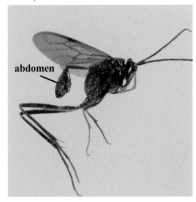

Fig. 595   Lateral view of a male ensign wasp. Note the small oval abdomen. Family Evaniidae

Suborder Apocrita
Ants, Bees, and Wasps

Family Pelecinidae* - *Pelecinid Wasps*   (Fig. 596)

Body shiny black and extremely long (20-70mm).
Abdomen very long and slender (about 4-5X the length of the head and the thorax combined).
* These characters apply only to the female.
Male resembles female, but abdomen club-shaped and not nearly as long.

Family Cynipidae - *Gall Wasps*   (Figs. 597-598)

Small (8mm or less).
Body usually shiny, and black or red in color.
Antennae thread-like, with 13-16 segments.
Venation reduced to the anterior portion of the wing.
Second abdominal segment makes up most of the abdomen.
Pronotum somewhat triangular when viewed in profile.
Dorsal surface of thorax may have a raised, O-shaped ridge.

Family Chrysididae - *Cuckoo Wasps*   (Figs. 599-601)

Small-medium (5-15mm).
Body metallic blue or green with many small pits in the surface.
Tip of the abdomen sometimes with several tooth-like projections.
Posterior corners of thorax pointed.
Hindwings with no closed cells.
Ventral portion of abdomen concave or hollowed out.

Family Apidae - *Honey Bees, Bumble Bees, & Carpenter Bees*

Medium-large (20-40mm).
Forewings with three submarginal cells.
Hind legs may have the tibiae and first tarsal segment enlarged,
flattened, and covered with hairs forming a 'pollen basket'.   (Figs. 605-606)
Tongue wide at the base and long and slender apically.
A 'stinger' may protrude from the tip of the abdomen.
Subantennal sutures originate at the centers of the antennal sockets.

Family Megachilidae - *Leafcutting Bees*   (Fig. 607)

Medium-sized (10-20mm).
Body shape distinctive.
Ventral surface of abdomen densely hairy forming a 'pollen brush'.
Forewings with two nearly equal-sized, adjacent submarginal cells.
Subantennal sutures originate from the outer margins of the antennal sockets.

Fig. 596   Female pelecinid wasp. Note the extremely long ovipositor. Family Pelecinidae

Fig. 597   Nearly mature gall wasp with half the gall removed.   Family Cynipidae *Photo JL Castner   Courtesy IFAS/UF*

Fig. 598   Lateral view of a gall wasp. Note the wide abdomen. Family Cynipidae

Fig. 599   Cuckoo wasp showing the typical metallic body color. Family Chrysididae

Fig. 600   Tip of a cuckoo wasp abdomen showing tooth-like projections. Family Chrysididae

Fig. 601   Lateral view of a cuckoo wasp in its rolled-up defensive posture. Family Chrysididae

Fig. 602   Honey bee. Family Apidae

Fig. 603   Bumble bee. Family Apidae

Fig. 604   Carpenter bee. Family Apidae

Fig. 605   Venation of a bumble bee. Family Apidae

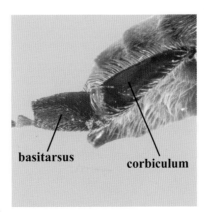

Fig. 606   Pollen basket (corbiculum) on the hind leg of a bee. Family Apidae

Fig. 607   Dorsal view of a leafcutting bee. Note the characteristic shape of the abdomen.   Family Megachilidae

Suborder Apocrita
Ants, Bees, and Wasps

Family Sphecidae - *Sphecid or Solitary Wasps*   (Figs. 608-610)

Pronotum short and collar-like, with posterior corners shaped like rounded lobes.
Lobes of pronotum do not reach or touch the tegulae.
Wings not folded longitudinally.
Inner margin of the eye not usually notched.
Many species (thread-waisted wasps) with an obvious petiole between the thorax and the abdomen.

Family Vespidae - *Social Wasps*   (Figs. 611-613)

Pronotum triangular-shaped when viewed in profile, and narrow at center when viewed dorsally.
Posterior corners of the pronotum pointed and reach or touch the tegulae.
Wings folded longitudinally.
Inner margin of eye strongly notched.

Family Pompilidae - *Spider Wasps*   (Figs. 614-616)

Pronotum rectangular or squarish when viewed in profile.
Posterior corners of pronotum rounded lobes that touch or are very near the tegulae.
Large rectangular sclerite (mesopleuron) above the middle leg with a transverse suture.
Legs long, with hind femora reaching or surpassing the tip of the abdomen.
Hind tibiae with two large spines at the apex.

Family Scoliidae - *Scoliid Wasps*   (Figs. 617-618)

Large (20-40mm), often hairy wasps.
Wingtips have many fine longitudinal 'wrinkles'.
Wings not folded longitudinally.
Eyes strongly notched.
Often black and yellow or black and red.

Family Tiphiidae - *Tiphiid Wasps*   (Fig. 619)

Tip of abdomen may have one or more short curved terminal spines.
Ventral area between middle legs with two posteriorly-pointing lobes.
Wingtips without fine longitudinal 'wrinkles'.

Fig. 608   Solitary wasp that parasitizes mole crickets.
Family Sphecidae

Fig. 609   Lateral view of a solitary wasp showing that the tegula does not touch the lobe of the pronotum.   Family Sphecidae

Fig. 610   Head of a solitary wasp. Note the inner margin of the eye is not notched.
Family Sphecidae

Fig. 611   Yellowjacket wasp.
Family Vespidae

Fig. 612   Lateral view of a social wasp showing that the tegula touches the point of the pronotum.   Family Vespidae

Fig. 613   Head of a social wasp. Note the inner margin of the eye is notched.
Family Vespidae

Fig. 614   Dorsal view of a spider wasp.
Family Pompilidae

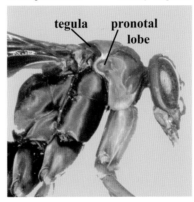

Fig. 615   Lateral view of a spider wasp showing the tegula is near or touching the pronotal lobes.   Family Pompilidae

Fig. 616   Lateral view of a spider wasp. Note the long hind legs.
Family Pompilidae

Fig. 617   Dorsal view of a scoliid wasp.
Family Scoliidae

Fig. 618   Tip of a scoliid wasp wing showing the fine longitudinal wrinkles.
Family Scoliidae

Fig. 619   Dorsal view of a tiphiid wasp.
Family Tiphiidae

Suborder Apocrita
Ants, Bees, and Wasps

Family Mutillidae - *Velvet Ants*

Wings present (males) or absent (females).
Typically brightly colored.
Moderately to densely hairy.
Abdomen with rings of hairs.

**Fig. 620　Female velvet ant.
Family Mutillidae**

**Fig. 621　Dorsal view of a male velvet ant.
Family Mutillidae**

**Fig. 622　Abdomen of a velvet ant showing the rings of hairs.
Family Mutillidae**

Family Formicidae - *Ants*

Wings present or absent.
One or two dorsal humps (nodes) present between the thorax and the main body of the abdomen.
Antennae elbowed.

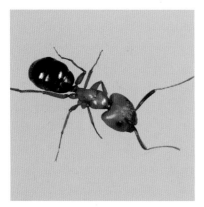

**Fig. 623　Carpenter ant.
Family Formicidae**

node

**Fig. 624　Lateral view of an ant showing a single dorsal node.
Family Formicidae**

**Fig. 625　Dorsal view of a twig-inhabiting ant.
Family Formicidae**

Social insects may sometimes form nests and colonies with a million or more individuals. When the species is both aggressive and venomous, like the yellowjackets (*Vespula squamosa*) pictured above, they can cause a legitimate threat to humans and animals. This giant yellowjacket nest was discovered on a golf course in south Florida. It was approximately eight feet high and four feet in diameter. It was estimated to contain over 250,000 adult yellowjacket wasps at the time of its removal.

Key to the Families of Hymenoptera

1A   Abdomen broadly joined to the thorax. Hindwing with three basal cells.
       (**Suborder Symphyta**) **2**

1B   Abdomen joined to the thorax by a thin stalk, or it narrows extensively to the point of
       attachment. Hindwing with no more than two basal cells. (**Suborder Apocrita**) **6**

2A   Front tibiae with one apical spur. **3**

2B   Front tibiae with two apical spurs. **4**

3A   Large (25-35mm). Stout dorsal spur at tip of abdomen. Pronotum much wider
       than long when viewed dorsally. .... **Siricidae**

3B   Small-medium (20mm or less). No dorsal spur at tip of abdomen. Pronotum
       approximately as wide as long when viewed dorsally. .... **Cephidae**

4A   Antennae with 13 or more segments, and serrate or pectinate. .... **Diprionidae**

4B   Antennae with less than 13 segments, and capitate or filiform. **5**

5A   Antennae with 7 segments or less and capitate. Medium-sized (15-25mm).   **Cimbicidae**

5B   Antennae usually with 9 segments and filiform. Size variable, but 20mm or less.
                                                             Tenthredinidae

6A   Abdomen extremely long (3X-5X length of head and thorax combined). Note: This
       character only applies to the female. .... **Pelecinidae**

6B   Abdomen not extremely long (less than or equal to 2X length of the head and thorax
       combined). **7**

7A   Hind femora greatly enlarged and serrate beneath. .... **Chalcididae**

7B   Hind femora not greatly enlarged. **8**

8A   Abdomen greatly reduced (smaller than thorax). Long slender stalk separates abdomen
       from thorax. .... **Evaniidae**

8B   Abdomen not greatly reduced (as big as or bigger than thorax). **9**

9A   Entire body metallic blue or green, with many small pits. Tip of abdomen usually with
       several tooth-like projections. Ventral portion of abdomen concave. .... **Chrysididae**

9B   Body not entirely metallic with tiny pits. Ventral portion of abdomen not concave. **10**

10A   Small (less than 10mm). Second abdominal segment makes up the majority of the abdomen. .... **Cynipidae**

10B   Size variable. Second abdominal segment does not make up the majority of the abdomen. **11**

11A   Wings folded longitudinally. Eyes strongly notched. .... **Vespidae**

11B   Wings not folded longitudinally. Eyes without notch or only weakly notched. **12**

12A   Wingtip with fine longitudinal striations or wrinkles present. .... **Scoliidae**

12B   Wingtip without fine longitudinal striations or wrinkles present. **13**

13A   Antennae with 16 or more segments. **14**

13B   Antennae with less than 16 segments. **15**

14A   Forewing with a large, somewhat trapezoidal-shaped cell below the stigma. A small, almost round cell (areolet) may be present beyond the stigma. Size variable (5-40mm).

                                                              Ichneumonidae

14B   Forewing with two closed cells below the stigma. No small, almost round closed cell present beyond stigma. Small (5-15mm). .... **Braconidae**

15A   Dorsal hump on first, or on first and second, abdominal segments when viewed laterally. Antennae strongly elbowed.   **Formicidae**

15B   No dorsal humps on first or second abdominal segments when viewed laterally. Antennae variable. **16**

16A   Body moderately to densely covered with brightly colored hairs. Abdomen with rings of hairs. .... **Mutillidae**

16B   Body bare or hairy, but without rings of hairs on abdomen. **17**

17A   Ventral portion of abdomen densely covered with hairs. .... **Megachilidae**

17B   Ventral portion of abdomen without dense covering of hairs. **18**

18A   Hind legs with tibiae and first tarsal segments flattened and bearing long hairs that act as a 'pollen basket'. .... **Apidae**

18B   Hind tibiae and first tarsal segments not flattened, may be covered with hairs, but not forming a 'pollen basket'. **19**

19A   Pronotum reaches the tegulae. Antennae often tightly curled. Apical spurs on tibiae very large and stout. .... **Pompilidae**

19B   Pronotum terminates in a rounded lobe that does not reach the tegulae. Antennae seldom tightly curled. Apical spurs on tibiae not overly large or stout. .... **Sphecidae**

Entomology Textbooks and Reference Books

Photographic Atlas Of Entomology And Guide To Insect Identification. 2000. Castner, J.L. Feline Press. Gainesville, FL.

A Field Guide To Insects. 1970. Borror, D.J. and R.E. White. Houghton Mifflin Co. Boston, MA.

Simon & Schuster's Guide To Insects. 1981. Arnett, R.H. Jr. and R.L. Jacques, Jr.. Simon & Schuster. New York, NY.

How To Know The Insects. 1978. Bland, R.G. and H.E. Jaques. William C. Brown, Co. Dubuque, IA.

An Introduction To The Study Of Insects. 1989. Borror, D.J., Triplehorn, C.A., and N.F. Johnson. Harcourt Brace College Publishers. Fort Worth, TX.

Fundamentals Of Entomology. 1997. Elzinga, R.J. Prentice Hall. Upper Saddle River, NJ.

Entomology. 1980. Gillott, C. Plenum Press. New York, NY.

A Textbook Of Entomology. 1991. Ross, H.H., Ross, C.A. and J.R.P. Ross. Krieger Publishing Co. Malabar, FL.

The Science Of Entomology. 1998. Romoser, W.S. and J.G. Stoffolano, Jr. McGraw-Hill Companies. Boston, MA.

The Insects - An Outline Of Entomology. 2000. Gullan, P.J. and P.S. Cranston. Blackwell Science Ltd. Oxford, U.K.

Introduction To Insect Biology And Diversity. 1998. Daly, H.V., Doyen, J.T. and A.H. Purcell, III. Oxford University Press, New York, NY.

The Insects - Structure And Function. 1998. Chapman, R.F. Cambridge University Press. Cambridge, U.K.

Insect Ecology. 1997. Price, P.W. John Wiley and Sons. New York, NY.

Immature Insects. 1981. Stehr, F.W. (Ed.) Kendall/Hunt Publishing Co. Dubuque, IA.

A Glossary Of Entomology. 1978. de la Torre-Bueno, J.R. New York Entomological Society. New York, N.Y.

Forensic Entomology. 2000. Byrd, J.H. and J.L. Castner (Eds.) CRC Press. Boca Raton, FL.

Forensic Insect Identification Cards. 2000. Castner, J.L. and J.H. Byrd. Feline Press. Gainesville, FL.

Amazon Insects - A Photo Guide. 2000. Castner, J.L. Feline Press. Gainesville, FL.

Glossary

abdomen - The posterior body region of an insect, typically containing eleven segments.

aedeagus - The insect penis or male genitalia.

ametabolous - The simplest form of development in insects, where the young or immature forms appear to be small versions of the adult.

anal loop - In dragonflies, a cluster of cells near the base of the hindwings located between the anal veins.

anal margin - The rear edge of the insect wing.

anal vein - One or several of the main longitudinal veins of the insect wing that are located nearest the anal margin or rear edge of the wing.

antenodal cross veins - In dragonflies, a series of short cross veins located between the wing base and the nodus that connect the front edge of the wing with the next longitudinal wing vein.

anterior - Referring to the front portion or head end of the body.

apical - Pertaining to that part of a segment or structure that is furthest away from the base.

apterous - Without wings. The condition of being wingless.

arculus - In dragonflies, a small cross vein near the base of the wing that connects the radius and cubitus veins.

arista - A large bristle or hair that projects from the third (last) antennal segment of flies in the Suborder Cyclorrhapha.

aristate - Antennal type characterized by a three-segmented antenna, where the third segment bears a protruding hair called an arista. Found on cyclorrhaphous flies.

basal cells - In bees and wasps, elongated closed cells that originate at the base of the wing.

bilateral symmetry - A type of symmetry possessed by arthropods, whereby a plane passed longitudinally through the center of the body will result in mirror images.

brace vein - In dragonflies, a diagonal cross vein that occurs below and is continuous with the proximal end of the stigma.

brachypterous - With short wings. The condition of having wings that typically do not reach the tip of the abdomen and are not capable of permitting flight.

calypters - In flies, two small lobes at the base of each wing near the anal margin.

campodeiform - A type of insect larva characterized by well developed thoracic legs, antennae, and cerci. Such larvae are typically elongate, somewhat flattened, and fast moving.

capitate - Antennal type characterized by having the tips enlarged into rounded knobs. Found on butterflies and owlflies, among others.

carapace - The hardened dorsal covering of the cephalothorax in certain arachnids.

caste - In social insects, a morphologically distinct group of individuals that performs certain functions within the nest or community.

cephalothorax - In arachnids, the anterior body region formed by the combined head and thorax.

cerci - Paired, feeler-like structures that occur at the tip of the abdomen.

chelicerae - In arachnids, the fangs or pincer-like mouthparts.

chrysalis - The pupal stage of a butterfly, typically suspended in some manner.

claspers - In butterflies, specialized clasping organs at the tip of the male's abdomen that are used to hold the female during mating.

clavate - Antennal type where the tip of the antenna is enlarged into a broad club. Found in carrion beetles, among others.

clavus - In true bugs, the triangular anal portion of the hemelytron that borders the scutellum.

closed cell - A cell on an insect wing that is entirely surrounded by either longitudinal veins or cross veins, with no part reaching any wing edge.

clypeus - An immovable rectangular flap on the insect face that occurs just below the frons and just above the labrum.

coarctate - A pupa that is surrounded by the hardened skin of the last larval instar.

complete metamorphosis - The most complex insect development pattern, characterized by passage through four distinct life stages (egg, larva, pupa, adult) and the internal development of the wings.

compound eye - An eye composed of individual visual units or facets that may number from several to several thousand.

corium - In true bugs, the thickened basal portion of the hemelytron.

costal margin - The front or leading edge of the insect wing.

couplet - A pair of clues or descriptions in a dichotomous key that is used to identify an organism to a certain taxonomic level. Only one of the descriptions in a given couplet should 'fit', and will lead the user to a new couplet and eventually the identification.

coxa - The basal segment of the insect leg found between the body and the trochanter.

cross vein - In the insect wing, a short vein that connects one longitudinal vein to another.

cubitus vein - In the insect wing, the main longitudinal vein that except for the anal vein is closest to the anal edge or margin.

cuneus - In true bugs, the thickened apical portion of the corium that is triangular in shape and bordered by the membranous wingtip and a suture that separates it from the rest of the corium.

cursorial - A type of insect leg that is long, slender, and used for running. Examples are found in roaches, tiger beetles, ground beetles, and others.

cuticle - The hard outer layer of the insect exoskeleton.

dichotomous key - A tool used to identify unknown organisms. Paired descriptions or clues called couplets guide the user through a series of steps that eventually lead to the name or identification.

distal - Referring to an area away from the body or originating at a point distant from the body.

dorsal - Referring to the top side or back area of the body.

ecdysis - The act of shedding the skin.

ectoparasite - An external parasite. For example, ticks, fleas, and lice are ectoparasites.

elateriform - A larval type characterized by a long cylindrical hardened body with short legs. An example is the 'wireworm' or larva of the click beetle (Family Elateridae).

elytron - In beetles, the forewing that is thickened and hardened, concealing the membranous hindwing beneath it except during flight.

epicranial suture - A forked suture with the appearance of an inverted 'Y' that comes straight down from the middle of the top of the insect head.

eruciform - A larval type that is a caterpillar (such as in butterflies and moths) or caterpillar-like (such as in sawflies). Characteristics include six thoracic legs and various pairs of fleshy abdominal prolegs that are used for locomotion.

exarate - A pupal type characterized by having the appendages (legs, antennae, wings) free from the body and easily distinguishable.

exoskeleton - The body wall of an insect which serves both protective and support functions.

exuvium - The empty shed skin of an insect.

femur - Third segment of the insect leg out from the body. Located between the trochanter and tibia, it is often the largest and most noticeable part of the leg.

filiform - Antennal type composed of a series of cylindrical or flattened segments which together present a thread-like appearance. Examples are seen in crickets, katydids, true bugs, and others.

flagellomere - Each individual segment that makes up the flagellum of an antenna.

flagellum - Name applied to all of the antennal segments collectively after the first two.

fontanelle - In termites, a light colored spot or depression on the top of the head between the eyes.

fossorial - A type of insect leg that is broad and shovel-like to facilitate digging through the ground. Heavily sclerotized claws may be present such as in mole crickets.

frenulum - In butterflies and moths, a spine or group of spines at the base of the hindwing that extend beneath the forewing and function to hold both wings together.

frons - The central area of the head or face that occurs between the two branches of the epicranial suture.

frontal suture - In flies, a suture on the face in the shape of an inverted 'U' or 'V'.

galeae - In butterflies and moths, those portions of the maxillae that have evolved to form the long coiled tongue.

gena - On the insect head, the 'cheek' or that area of the side of the head below the eyes.

geniculate - Antennal type that is characterized by being abruptly bent or elbowed. Examples are found on ants, weevils, and bees.

gradual metamorphosis - Insect development characterized by young that resemble the adults except for the presence of wings. The immatures are called nymphs and develop the wings gradually as external pads that get increasingly larger as they mature.

haustellum - In flies, the distal portion of the sponging mouthparts that bears the labella.

haltere - In flies, the small knobbed structure on each side of the metathorax that replaces the hindwings.

head - The most anterior insect body region or tagma, which bears the mouthparts and many of the sensory structures such as antennae and compound eyes.

hemelytron - In true bugs, the forewing that has the basal portion thick and leathery while the wingtip remains membranous.

hemimetabolous development - Insect development characterized by young that resemble the adults except for the presence of wings. The immatures are called nymphs and develop the wings gradually as external pads that get increasingly larger as they mature.

holometabolous development - Insect development characterized by the insect passing through four distinct life stages (egg, larva, pupa, adult) and the internal development of the wings. Also referred to as complete metamorphosis.

hypermetamorphosis - A form of holometabolous development in parasitic insects where larval development includes more than one form. Typically, the first instar in a mobile free-living insect while the later larval instars are immobile and parasitic.

hypopharynx - A fleshy knob on the upper surface of the labium, which constitutes the 'tongue' in those insects with chewing mouthparts.

hypopleuron - In flies, a sclerite on the thorax just above the hind coxa.

instar - Refers to the insect itself at any given life stage. For example, a third instar larva is an immature insect that has already shed its skin twice as a larva.

jugum - In butterflies and moths, a lobe at the base of the forewing that extends over the hindwing and serves to hold the two wings together.

key - Descriptions used in determining the identification of an organism. (See **dichotomous key**.)

labella - In flies, two fleshy lobes at the tip of the haustellum that function to draw up liquid to the food channel.

labial palps - Three-segmented appendages that are found on each side of the labium.

labium - A flap-like structure located below the maxillae and that can be thought of as a 'lower lip'.

labrum - The flap-like, movable structure that occurs just below the clypeus and can be thought of as the 'upper lip'.

lamellate - Antennal type that is clubbed, but where the terminal segments are enlarged parallel plates that stick out perpendicular to the rest of the antennae.

larva - In insects that undergo complete metamorphosis, the life stage that occurs after the egg.

lateral ocelli - Also called stemmata, they are simple eyes or photoreceptors on the head of some larval insects.

longitudinal veins - The long veins in the insect wing that originate near the base and run nearly parallel to the costal margin or front edge.

mandibles - Paired structures, one on each side of the mouth, that are used to tear, bite, grind, and chew food. In some insects they have become highly modified.

marginal cells - Closed cells that border the front edge of the forewing from the stigma to the wingtip.

maxillae - Paired structures, one on each side of the mouth and below the mandibles, that contain both a fleshy and a sharpened area. They are used in feeding and may be highly modified in some groups.

maxillary palpi - Five-segmented appendages, one each originating at the base of each maxilla.

media vein - In the insect wing, the fourth of the longitudinal veins located between the radius and the cubitus veins.

membrane - In true bugs, the wigtip or membranous portion of the hemelytron or forewing.

mesothorax - The second or middle thoracic segment which bears the forewings (when present).

metamorphosis - The process of undergoing physical changes from one life stage to the next.

metathorax - The third or most posterior thoracic segment which bears the hindwings (when present).

molting - The process by which an insect develops a new skin and sheds the old.

moniliform - Antennal type characterized by a series of bead-like or rounded segments. Such antennae are found on termites and on some beetles.

naiad - Older term that describes the aquatic immature forms of insects that undergo hemimetabolous development, and which are usually characterized by some type of gill structures. Refers to the nymphs of dragonflies, damselflies, stoneflies, and mayflies.

natatorial - A type of insect leg that is used for swimming and that is characterized by being flattened or bearing long hairs that can be used as oars to propel the insect through water.

node - In ants, a dorsal hump that is located on the first and second abdominal segments.

nodus - In dragonflies and damselflies, a small heavy cross vein near the middle of the front edge of the wing.

notum - The top portion of a thoracic segment.

obtect - A pupal type characterized by having the appendages tightly glued to the body so that the pupal surface appears smooth and without projections. Such pupae are found in the butterflies and moths, as well as in some flies.

ocellus - A simple eye consisting of a single facet that is used to detect differences in light intensity and found on the head of some adult insects.

ommatidia - The facets or light-sensing visual elements that make up the compound eye.

open cell - In the insect wing, a cell that has any part of its perimeter formed by a wing edge.

opisthosoma - The abdomen or posterior body region in some arachnids.

ovipositor - An egg-laying device, variable in form and size, that is located at the tip of the female abdomen.

paurometabolous development - Older term used to refer to hemimetabolous insect development among the orders where the immature stages are not aquatic.

pectinate - Antennal type that is comb-like, with lateral processes that stick out at regular intervals.

pedicel - The second segment of an insect antenna. In arachnids, a stalk that connects the abdomen to the cephalothorax.

pedipalps - In arachnids, a pair of appendages associated with the mouth. They may be modified to be claw-like or antenna-like.

petiole - In wasps, the slender stalk-like portion of the first abdominal segment.

pleural - Pertaining to the sides or lateral surfaces of the body.

pleural membrane - The membranous side areas of the abdomen, between the terga and sterna.

pleuron - On the thorax, the side area of each segment located between the notum and sternum.

plumose - Antennal type that is feather-like or with whorls of hair. Examples are found on giant silk moths and on male mosquitoes.

posterior - Pertaining to the rear or tail end of the body.

postscutellum - In flies, the area just behind or below the scutellum.

proboscis - An extended mouth structure such as a beak or long tongue.

proleg - One of the fleshy abdominal legs used for locomotion in caterpillars and other eruciform larvae.

propodeum - In wasps, the first abdominal segment which is fused to the thorax.

prosoma - In arachnids, the cephalothorax or anterior body region.

prothorax - The first or most anterior segment of the thorax, located directly behind the head.

proximal - Pertaining to something that is close to or originating near the body.

pterothorax - The second and third segments of the thorax together. The part of the thorax bearing the wings.

pulvilli - In flies, pads or lobes located beneath the tarsal claws.

pupa - In insects that undergo complete metamorphosis, the typically immobile life stage that follows the larva and precedes the adult.

puparium - In some flies, the hardened larval skin that surrounds the pupa.

radius vein - In the insect wing, the longitudinal vein that occurs after the subcosta and before the cubitus.

raptorial - A type of insect leg that is used for catching prey, typically characterized by a series of sharp spines. Examples are found on mantises and assassin bugs, among others.

rostrum - In flies with sponging mouthparts, the basal portion of the proboscis that attaches to the head.

saltatorial - A type of insect leg that is used for jumping, typically characterized by an enlarged femur.

scale - In termites, the wing remnant or stub that remains attached to the thorax after the wings have been shed. In the Homoptera, families of insects that are characterized by a waxy covering that often completely conceals the body.

scansorial - A type of insect leg that has evolved for clinging to and climbing among hair. Such legs are usually characterized by specialized claws and are found on ectoparasites.

scape - The first or basal segment of an insect antenna, which is sometimes enlarged.

scarabaeiform - A type of insect larva that is grub-like and only capable of limited movement.

sclerite - A hardened plate, many of which typically make up a large portion of an insect's body surface.

sclerotin - The chemical material responsible for the hard parts of the insect's exoskeleton.

sclerotization - The process of hardening of parts of the insect exoskeleton.

scutellum - In true bugs, flies, and beetles, a triangular-shaped sclerite that is found on the dorsum directly behind the thorax and between the base of the wings.

serrate - Antennal type that is saw-toothed in appearance and composed of roughly triangular segments.

setaceous - Antennal type that is bristle-like in appearance. Typically slender and tapering to a tip. Examples are found on dragonflies, damselflies, and cicadas.

spinnerets - In some arachnids, small finger-like structures that produce silk and are located on the underside of the abdomen near the anus.

spiracle - A hole or valve in the insect body wall that is the external opening and connection to the respiratory system.

stadium - The period of time an insect spends in any particular life stage.

stemmata - Also called lateral ocelli, these are simple eyes found on the heads of certain larval insects and on the adults of some primitive insects.

sternum - Term applied to the bottom portion of each thoracic segment, and the bottom part or sclerite of each abdominal sgment.

stridulatory organ - A sound-producing mechanism that usually involves one structure rubbing against another.

stylate - Antennal type whose tip terminates in a long, slender point called a style. An example is found on the robber flies (Family Asilidae).

style - The slender terminal portion of a stylate antenna, which may be hair-like or thicker and more substantial.

stylets - In the piercing-sucking mouthtype, extended sharpened structures that are used for piercing.

sulcate - A mouthtype characterized by grooved, sickle-like jaws.

sulcus - The groove in the jaws of a sulcate mouthtype that channels the juice or body fluid of the prey to the mouth.

tagma - An insect body region formed by the fusion of various segments. The three insect tagmata are the head, thorax, and abdomen.

tarsal formula - In beetles, pertaining to the number of tarsal segments or tarsomeres found on each of the pair of legs beginning with the prothoracic pair. A tarsal formula of 5-4-3 signifies that there are five tarsal segments on the front pair of legs, four on the middle pair, and three on the hind pair.

tarsomere - A segment of the insect tarsus.

tarsus - The terminal portion of the insect leg composed of segments called tarsomeres, and often with a claw.

tegmen - In Orthoptera, the leathery forewing which covers and conceals the membranous hindwing. Tegmina often serve to protect while the membranous wings are actually used for flight.

tegula - In wasps and bees, a plate-like sclerite that covers the base of the forewing.

telson - In arachnids, the region of the abdomen that is posterior to the anus.

tentorium - A system of braces inside the insect head that serves to reinforce and provides points of attachment for muscles.

tergum - The top part or sclerite of a segment of the insect abdomen.

thorax - The middle insect body region or tagma which bears the legs and wings.

tibia - Fourth segment of the insect leg distal from the body, which is located after the femur and before the tarsus.

triangle - In dragonflies, a closed triangular-shaped cell near the base of the wings that is used in identification.

trochanter - Second segment of the insect leg distal from the body, which is located after the coxa and before the femur.

tympanum - The insect ear, typically characterized by an oval depression or flattened area on the body.

ventral - Pertaining to the bottom side of the body.

vermiform - A type of insect larva that is worm-like or maggot-like in appearance. Such larvae are legless and may or may not have a distinctive head.

vertex - The top of the head, and point where the epicranial suture originates.

vestigial - Pertaining to something that is lacking or greatly reduced. For example, some insects that only live a very short period of time as adults do not feed. They have no mouthparts, or may be said to have vestigial mouthparts.

About the Author

Dr. James L. Castner is an entomologist and tropical biologist who writes, photographs, edits, and publishes books. His academic background includes a doctoral degree in entomology with a minor in botany. Following the receipt of his doctorate, he worked as a research biologist and as a Scientific Photographer for a major university. Recently, to better understand the training of educators, Dr. Castner has added the education courses necessary for certification to teach Science and Spanish at the secondary school level. This academic foundation, combined with the experience of over 20 years of travel throughout Latin America and the Amazon Basin, have provided the expertise to teach and write on a variety of topics for diverse audiences including children, college students, and educators.

In 1990, Dr. Castner established the company Feline Press to publish his first book titled *Rainforests: A Guide To Research And Tourist Facilities At Selected Tropical Forest Sites In Central And South America*. The success of this venture led him to believe that there was a genuine need for small independent publishers that could publish titles on very specific subjects which are typically ignored by large 'professional' publishers. Since then, he has identified and filled vacant 'literary niches' with his own books. He left his academic position in order to follow this pursuit full time.

In addition to this book and the title above, Dr. Castner has also co-written or co-edited the following titles: 1) *Field Guide To Medicinal And Useful Plants Of The Upper Amazon* 2) *Amazon Insects - A Photo Guide* 3) *The Amazon Rainforest - An Exploration Of Countries, Cultures, And Creatures - A Learning Center For Secondary School Students* 4) *Explorama's Amazon - A Journey Through The Rainforest Of Peru* 5) *Forensic Insect Identification Cards* and 6) *Forensic Entomology - The Utility Of Arthropods In Legal Investigations*. Dr. Castner is currently working on a series of children's books titled *Deep In The Amazon* that are tentatively scheduled for publication by Marshall Cavendish Publishers in 2001. Following this, he has plans for many more heavily illustrated works on entomology and the other life sciences.

During his career as a writer and photographer, Dr. Castner has had articles or photos published in *National Geographic, Natural History, International Wildlife, Ranger Rick, National Geographic WORLD, GEO, GeoMundo,* and *Kids Discover*. His photos appear in almost every college-level biology textbook currently in use. His favorite photographic subjects are examples of insect defense mechanisms, including camouflage and mimicry. The latter includes many unique shots of neotropical leaf-mimicking katydids and their displays.

Although Dr. Castner intends to continue writing, he wishes to increase his interaction with students and educators. He currently seeks a position at a small university where he might teach entomology, tropical biology, and a variety of other courses related to the life sciences, forensics, and photography. He firmly believes in field-oriented, hands-on courses where students can learn and have fun at the same time.

Photographic Atlas Of Entomology
And
Guide To Insect Identification

Available for the first time, a learning aid created especially with entomology students in mind. Dr. James L. Castner, entomologist and professional photographer, has combined his teaching experience with his photographic skills to create *Photographic Atlas Of Entomology And Guide To Insect Identification*. Designed specifically for college students of General Entomology and Insect Taxonomy classes, this is the first thorough, photo-oriented entomological guide of its kind.

Dr. Castner has provided over 600 color photos that illustrate the structural characters and anatomical features of the major insect families and arthropod groups discussed in entomology courses. In addition to photographic coverage of approximately 190 arthropod taxa (including 30 insect and 8 arachnid orders), explanations and illustrations of external anatomy and insect development are also provided. Dichotomous keys to the family level are included for the major orders, as well as a glossary of specialized terms.

Entomology professors and lab instructors will appreciate that this guide helps to provide a degree of consistency in the coverage and information available from one lab section to the next. Students will appreciate that the main characters used in identification are already listed for them, allowing more time to be spent in examining actual specimens. This work permits students to take an insect reference collection home with them!

Title:	*Photographic Atlas Of Entomology And Guide To Insect Identification*		
Author:	**James L. Castner**	Size:	**8.5" X 11"**
Publisher:	**Feline Press**	Pages:	**192**
ISBN:	**0-9625150-4-3**	Photos:	**670**
Price:	**$35**	Binding:	**Spiral**

Bookstores and quantity orders receive traditional discounts.

For single copies, please add $5 for shipping via Priority Mail within the United States. Please make checks or money orders payable to Feline Press.
No credit cards are accepted.

Please send orders to:

Feline Press
P.O. Box 357219
Gainesville, FL 32635 USA

Questions may be addressed to: jlcastner@biologicalphotography.com
jlcastner@aol.com

174